Energy and Environment in the Transition Economies

Between Cold War and Global Warming

William Chandler

Routledge
Taylor & Francis Group
New York London

First published 2000 by Westview Press

Published 2018 by Routledge
605 Third Avenue, New York, NY 10017
2 Park Square, Milton Park, Abingdon, Oxon OX14 4RN

Routledge is an imprint of the Taylor & Francis Group, an informa business

Library of Congress Cataloging-in-Publication Data
Chandler, William U., 1950–
    Energy and environment in the transition economies : between Cold War and global warming / William Chandler.
        p.   cm.
    Includes bibliographical references and index.
    ISBN 0-8133-3812-3
    1. Energy policy—Europe, Eastern.   2. Energy policy—Former Soviet republics.   3. Environmental policy—Europe, Eastern.   4. Environmental policy—Former Soviet republics.   5. Energy industries—Europe, Eastern.   6. Energy industries—Former Soviet republics.   I. Title.

HD9502.E832 C5   2000
333.79'0947—dc21

                                                                                    00-039880

ISBN 13: 978-0-8133-3812-5 (pbk)

Contents

Illustrations

Figures

Boxes

Maps

Acronyms

AIOC	Azerbaijan International Operating Company
CENEf	Center for Energy Efficiency
CPC	Caspian Pipeline Consortium
EHDP	Enterprise Housing Divestiture Program
EBRD	European Bank for Reconstruction and Development
ESCO	energy service company
FCCC	Framework Convention on Climate Change
FEWE	Polish Foundation for Energy Efficiency
G-7	Group of Seven (leading industrial nations)
GEF	Global Environment Facility
GDP	Gross Domestic Product
IEA	International Energy Agency
IFC	International Finance Corporation
IGCC	integrated, gasification of coal, and combined cycle
IRP	Integrated Resource Planning
IRR	internal rate of return
MPC&A	Materials Protection, Control, and Accounting Methods
PBR	Polski Bank Rozwoju (Polish Development Bank)
REMA	Russian Energy Managers Association
SEVEn	Czech Energy Efficiency Center
USEC	U.S. Enrichment Corporation

Preface

The problem with writing a book about the transition economies is that they keep changing. It occurs to me that the very idea of a finite transition to market democracy is an illusion. The struggle never ends, even for those of us who live in 200-plus–year-old democracies. I hope, at least, that the reader will excuse me for pausing now to publish this book before the story it tells has ended.

My understanding of change in the transition economies has been shaped by the instruction, help, and encouragement provided by many colleagues, several of them now close friends. Their courage, labor, friendship, and sacrifice have helped me to realize the value and the cost of reform. Individuals whose blood, sweat, and tears show in these pages include, in alphabetical order: Igor Bashmakov, Zbigniew Bochniarz, Estelle Chandler, Jae Edmonds, Bill Eichbaum, Zdravko Genchev, Jack Gibbons, Holly Gwin, Andrzej Kassenberg, Marc Ledbetter, Jaroslav Marousek, Grazyna Michalik, Sławomir Pasierb, Lydia Popova, Mykhailo Raptsun, Tom Secrest, Paul Schwengels, Glen Skovholt, Gerry Stokes, Bill Wiley, and Zhou Dadi.

A few world-class clients have directly and indirectly provided the foundation that supported everything else: Beth Arner, Jon Blythe, Henryka Bochniarz, Ron Bowes, Rick Bradley, Brian Castelli, Jon Elkind, Christine Ervin, John Firor, Robert Ichord, Wil Kohl, Mykhailo Kovalko, Dina Kruger, John Millhone, Erik Pearson, Bill Pennell, Bill Riley, Annie Roncerel, Dan Stein, Dennis Tirpak, Dennis Whittle, Dana Younger, and Zhang Ji-Qiang. Institutionally, they represent the Battelle Memorial Institute, the Global Environment Facility, the Johns Hopkins University School of Advanced International Studies, the International Finance Corporation, the W. Alton Jones Foundation, the John D. and Catherine T. MacArthur Foundation, the C. S. Mott Foundation, the National Center for Atmospheric Research, the National Security Council, the Pacific Northwest National Laboratory, the United Nations Development Program, the U.S. Agency for International Development, the U.S. Department of Energy, the U.S. Environmental Protection Agency, the U.S. Trade Development Agency, the World Bank, the White House Office of Science and Technology Policy, and the World Wildlife Fund.

Close colleagues who have advised, corrected, and helped me include: Matthew Auer, Meredydd Evans, Jeffrey Logan, Susan Legro, Karen King, Victoria Larue Garzillo, Adam Guła, Ewaryst Hille, Inna Gritsevich, Stan Kolar, Susan Legro, Steve Mtadineo, John Parker, Jerzsy Piszczek, Ilya Popov, Dee Sutton, Meredith Tirpak, Paulette Killgo Wright, and Vladimir Zhuze. Scientists, businesspeople, economists, researchers, and leaders who have shared their knowledge with me include: Rick Adcock, Fegan Aliyev, Chris Andreoni, Mikhail Berner, Eric Bielke, Simeon Bozhanov, Silviu Brucan, Vitaly Bushuev, Michael Congdon, Robert Ebel, Greg Kats, June Koch, Avtandil Koridze, Mark Levine, Yuri Matrosov, Lisa McNeilly, Jeff Malakoff, Bedreich Moldan, Matt O'Neill, Astghine Pasoyan, Alex Sich, Glen Skovholt, Rob Socolow, Stanisław Sitnicki, Einar Tresselt, Frank von Hippel, Rob Watson, Jay Wertenberger, and Jirka Zeman. Persons whom I have not met but whose work has been vital—János Kornai comes immediately to mind—are too numerous to mention individually. Scholars at Johns Hopkins who provided detailed comments include Eric Busch, Michael Rucker, Michael Trainor, Nuzhnet Orekli, Stian Christensen, and Robert Heilman.

Colleagues at Westview, Karl Yambert, Lisa Wigutoff, and Sharon DeJohn, ably facilitated the editorial and production processes.

Many thanks to each of you.

William Chandler
Washington, D.C.

Central and Eastern Europe

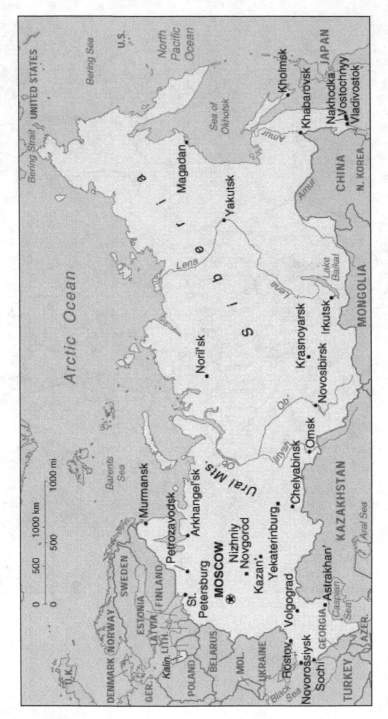

Russia

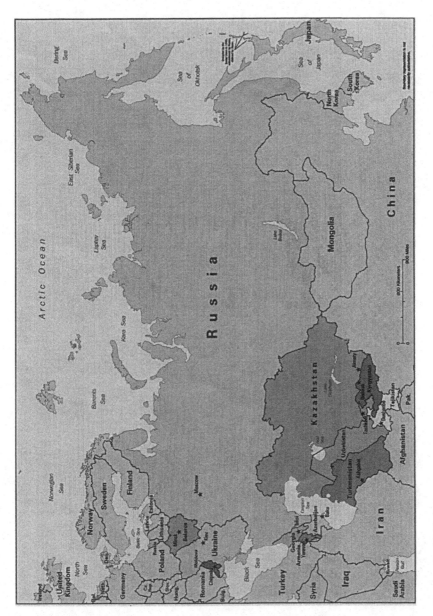

Former Soviet Union

1

Energy and Change

Antonio Gramsci wrote that "The problem consists precisely of the fact that the old is dying and the new cannot yet be born: in this interregnum, a great variety of morbid symptoms appears."[1] His metaphor, written around 1930 in a letter from prison, described the death of capitalism and the birth of communism; this Italian revolutionary would be appalled that this description now characterizes the opposite. Modern writers notwithstanding, the 1991 collapse of the Soviet empire was not the end of history.[2] The future remains up for grabs in more than two dozen nations in a transition zone stretching from the Baltic to the Bering Sea. The future of this region, its economy and environment, will be shaped in the kilns of energy policy.

The international community places priority on energy in the transition economies because security, trade, and environmental concerns compel it. Within what was called the Soviet Bloc, nuclear facilities produce the world's least secure stockpile of nuclear weapons-grade material. Multinational companies seek the richest remaining petroleum prizes in the largest wilderness and most unstable regimes. The Russian coal industry, the world's deadliest except for China's, kills more people each year than the Chernobyl disaster did in ten years. The world's least efficient nations produce more greenhouse gas emissions than all others except the United States and China, and perhaps will offer the cheapest emissions mitigation opportunities. Energy technology and policy created these conditions, and the energy industry continues as the least-reformed sector in the post-Communist world.

The economies in transition encompass over 400 million people living in the regions of Central and Eastern Europe, Russia, and Central Asia (see Table 1.1). Snapshots of this super-region suggest the extent of its energy waste and resulting pollution. The region's cities in the early 1990s were a bleak industrial landscape. Children played in soot-choked residential streets built alongside giant steel and coking operations, chemical

---

BOX 1.1   THE ECONOMIES IN TRANSITION

The *transition economies* include 26 nations: 15 countries created from the
Soviet Union–Armenia, *Azerbaijan*, Belarus, Estonia, Georgia, *Kazakhstan*,
Kyrgyzstan, Latvia, Lithuania, Moldova, *Russia*, Tajikistan, Turkmenistan,
Ukraine, and Uzbekistan; 6 European countries of the Council on Mutual
Economic Assistance–*Bulgaria*, the *Czech Republic*, *Hungary*, *Poland*,
Romania, Slovakia; and 5 nations created from shattered Yugoslavia.

This book focuses on the nations indicated in italics, which with a couple of
exceptions have populations greater than 10 million people.

---

plants, and unfiltered power stations. In some areas, the skies were filled
with particulates, carcinogenic benzopyrene, acid mist, arsenic, and antimony.[3] Rivers carried refinery waste and sewage; some carried radiation.
The environmental situation has improved but often as a result of economic depression, and human health conditions have deteriorated dramatically in some areas.

It is possible to understand change in this array of economies by con-
centrating on the larger countries and on the strong connections between
energy and economy and energy and environment in those countries.
Just four countries—Russia, Ukraine, Poland, and the Czech Republic—
account for three-quarters of regional energy consumption. Just two
countries—Russia and Poland—account for three-quarters of regional
energy production (see Table 1.2). Consumption in the region requires
only three-quarters of production, but Russia accounts for most of the
surplus. None of this means the other countries are unimportant: Reform
experience in the Czech Republic and Hungary provides valuable
lessons for the other transition nations; the Central Asian nations possess
oil resources to rival the Middle East; tiny Lithuania operates two of the
world's twenty most dangerous reactors; the Baltics, the southeastern
European nations of Bulgaria and Romania, may be key to geopolitical
conflict resolution, as may Armenia, Georgia, and Chechnya; and Central
Asian nations figure in western policy toward Iran, affecting Caspian oil
development priority policies.[4] Policy makers'—and their clients'—survival depends on the ability to identify priorities in this swirling blizzard
of change.

At Stake

Energy and environment will matter to transition nation governments
and businesses and their western partners for decades to come. Three en-

TABLE 1.1   Energy Use in Transition Economies, 1999

Country	Population (millions)	Energy Use (exajoules)	Self-Sufficiency[a] (percent)
Russia	147	26.0	165
Ukraine	51	5.8	53
Poland	39	4.0	86
Uzbekistan	24	2.3	119
Kazakhstan	17	1.5	167
Romania	23	1.5	72
Czech Republic	10	1.5	77
Belarus	10	1.0	13
Hungary	10	1.0	61
Bulgaria	8	0.8	49
Azerbaijan	8	0.5	156
Turkmenistan	4	0.6	148
Total Region[b,c]	407	48.2	130

NOTES: a) "Self-sufficiency" means domestic production as a percent of domestic commercial energy demand. A "self-sufficient" nation may be dependent on imported oil or gas. b) Includes all transition economies. c) For comparison, Germany and the United States have populations of 82 and 274 million, and energy use of 14.1 and 95 exajoules, respectively.

SOURCE: Derived from British Petroleum, *Statistical Review of World Energy, 2000* and United Nations (population).

ergy issues form a core of concern, and these will remain whether the region completes the economic transition or turns back. First, civilian application of military nuclear materials multiplies the threat of terrorism. Material enough to make several thousand nuclear bombs will be processed through the Russian nuclear industry for use as reactor fuel. Russian institutions are unprepared for that Herculean responsibility. The region, no less than the west, could be menaced by diversion of nuclear materials and the blackmail and terror that would go with it. In addition, although nuclear power provides a relatively small share of the region's energy, some thirty nuclear power reactors continue to operate with flawed design and inadequate containment (see Table 1.3). Russia and Ukraine, which operate the most unsafe reactors, have alternative power sources available but do not switch to them for political and secu-

TABLE 1.2   Energy Supply by Type, 1999 (percent of total production)

	Oil	Gas	Coal	Nuclear	Hydro	Total
Azerbaijan	55	44	0	0	1	100
Belarus	42	57	2	0	0	100
Bulgaria	25	13	37	24	1	100
Czech Republic	22	18	51	9	1	100
Hungary	30	40	15	15	0	100
Kazakhstan	29	15	54	<1	2	100
Poland	21	11	68	0	<1	100
Romania	32	43	17	4	4	100
Russia	21	55	17	5	2	100
Slovakia	21	38	28	17	2	100
Turkmenistan	30	70	0	0	0	100
Ukraine	11	46	28	15	1	100
Uzbekistan	14	82	3	0	1	100
Other FSU	31	37	6	14	11	100
World average	40	24	26	7	3	100

SOURCE: Derived from British Petroleum, *Statistical Review of World Energy, 2000.*

rity reasons. The nuclear industry includes not only power plants and their workers but a large infrastructure of engineering, manufacturing, and processing facilities involving hundreds of thousands of workers who see the prospect of commercial power as a chance to maintain their position.

Second, Russian and Caspian oil resources affect world markets, western energy security, and regional stability. The Caspian's vast oil discoveries have become the prize not just in a struggle for energy and money but in a geopolitical game that alternately pits U.S. and European interests with and against Iran, the Middle East, Russia, China, Turkey, Pakistan, and India. The decrease in Russian annual oil production during the 1990s was greater than the amount the United States now produces. Russia still ranks third in world oil output, after Saudi Arabia and the United States, and exports much of this wealth. Indeed, the collapse in world oil prices pulled a last leg from under Russia's currency, markets, and government in August 1998. That price slump—and the sharp oil price increase in 2000—also affected development of the Caspian region's vast oil resources. More important, there is trouble in every direction from the Caspian to world markets.

TABLE 1.3 Selected Indicators of Economic Reform in the Transition Economies[a]

Country	GDP Index (1991 = 100)	Government Deficit (% of GDP)	Consumer Prices (% per year)	Private Sector (%)c	Unemployment (%)	Life Expectancy[d] Male (years at birth)	Life Expectancy[d] Female (years at birth)
Azerbaijan	53	-2.7	-9	~20	1.1[b]	63.4	73.5
Belarus	84	-1.4[b]	294	~15	3.8[b]	n.a.	n.a.
Bulgaria	83	-1.4[b]	n.a.	~35	13.2	67.1	74.9
Czech Republic	102	-1.7	2	~70	8.4	70.0	76.9
Hungary	116	-0.3	10	~60	10.5	65.3	74.5
Kazakhstan	70	-3.4[b]	8	~25	3.9[b]	n.a.	n.a.
Poland	148	-2.3	7	~60	11.9	67.6	76.4
Romania	91	-2.6	46	~40	11.8	65.7[b]	73.4[b]
Russia	59	-5.7	86	~60	14.2[b]	58.3	71.7
Ukraine	40	-4.1	23	~35	n.a.	n.a.	n.a.

NOTES: a) All data for 1999, except where noted; b) 1997–98 average; c) 1997; d) 1995 one-year data.

SOURCES: International Monetary Fund, *World Economic Outlook* (Washington, D.C.: IMF, 2000); *Business Central Europe,* "Statistics," www.BCEMag.com.

Third, climate change has become a global challenge commensurate with the Cold War. The transition economies offer the world's largest and cheapest near-term opportunities for curbing greenhouse gas emissions and could do well in any emissions trading or joint implementation scheme. The region has seen breathtaking reductions in greenhouse gas emissions. Between 1990 and 2000, regional carbon dioxide emissions fell from 21 percent of the global energy-related total to just 13 percent, due mainly to economic decline. Central Europe has proven that economic recovery can mean reduced energy intensity and lower rates of carbon emissions. Whether the region follows Poland, the Czech Republic, and Hungary's example will decisively affect international climate goals.

How these economies respond to climate change will also affect local air and water pollution, which cause unnecessary early death and suffering and threaten the world's largest wilderness. Environmental quality suffers acutely from energy development. Spilled and discharged oil damages sensitive environments. Western Siberia, the world's largest wetland, still produces most of the region's oil. Three times the size of Texas, a third of Western Siberia is flooded or frozen most of the time. Oil is sloughed off in pumping operations, spilled when pipelines break or pumping stations fail and relief valves trip, oozing into the marshes, lakes, and rivers, enmeshing aquatic life in toxic hydrocarbon chains and rings. Western firms could improve or worsen this prospect. Oil shipments over the Black Sea could triple in the next fifteen years, and with it the risk of catastrophic oil spills in that shallow sea and along the narrow Bosporus. Western governments provide large sums of subsidized credit to firms exploiting these natural resources, but not to conserve and protect those resources for the future.

The region has the potential to create cross-cutting solutions to its energy problems. Over the next two decades, greater efficiency could cut the rate of energy used per unit of output by one-quarter to one-half, but lack of political and financial infrastructure impedes cost-effective solutions. Fuel switching to abundant natural gas would cut carbon dioxide emissions by half compared to coal. Indeed, gas can be transported long distances and delivered at competitive costs, and it is exceedingly abundant, with decades of supply available at current prices. Moreover, linking China and South Asia to Siberian and Central Asian gas supplies could also improve efficiency and reduce global as well as local pollution. Russia and Ukraine have extensive, sophisticated gas turbine manufacturing capacity, but their machines still lag in efficiency behind the west by a decade. If the future depends on natural gas, market reform and technology improvement must be accelerated. Gas customers burn a premium fuel in grossly inefficient systems and, more often than not, fail to pay for what they use. It is of no small concern that one company, Russia's Gazprom, dominates this vital resource.

After the Fall

Nothing physical distinguished the planned economies better than energy intensity (see Box 1.2): not energy supply, which varied from heavy reliance on coal in Poland and the Czech Republic to heavy reliance on gas in Russia and Ukraine; not even living standards, which between Central Europe and Central Asia ranged from First to Third World levels. But the technology and practice of energy use in factories, apartments, and cars from Warsaw to Vladivostok conformed to a striking wastefulness. Planned economies tend to be more and market economies less energy intensive, which is defined here as energy consumed per unit of economic output.[5] Although western energy use may be physically and environmentally excessive, its economic efficiency is far higher than in the transition economies. Higher energy prices, tighter fiscal budgets, and greater competitive pressure should in theory reduce energy waste and the costs associated with it. Reform, or the lack thereof, should be mirrored in the changing energy intensity of the transition economies over time, answering the questions of what has been and what remains to be done.

BOX 1.2   ENERGY INTENSITY: AN ILLUSTRATION

Russia:
- GDP = $4,000 GDP per capita
- Energy use = 170 GJ per capita
- Energy intensity = 177 GJ per capita/$4,000 GDP per capita = **42 MJ/$ GDP**

United States:
- GDP = $25,000 GDP per capita
- Energy use = 334 GJ per capita
- Energy intensity = 365 GJ per capita/$25,000 GDP per capita = **13 MJ/$ GDP**

NOTE: EJ means exajoules, or 10^{18} Joules. MJ means megajoule, or 10^6 Joules. EJ and MJ are roughly equal to a quadrillion and a thousand BTU, respectively. GDP means Gross Domestic Product. GDP in Russia might be as high as $6,000 per capita. If so, Russian energy intensity would be about 30 MJ.
SOURCE: Author's estimates.

If one could judge by primary energy demand figures alone, the answers would be clear. Energy consumption in the transition economies has fallen by 25 exajoules since 1990 (see Figure 1.1). That decline equals two-and-one-half times total demand for all of Africa. Energy demand in the region fell by one-third, with Ukraine falling hardest, tumbling 40 percent since 1990. In Ukraine, as well as Hungary and Romania, coal use fell by one-half despite dependence on oil and gas imports. Gas consumption in the Czech Republic was up 50 percent, partly as a result of a government clean air campaign. Oil consumption was down over one-fifth in Central Europe and by one-half in the former Soviet Union. Nuclear power use increased relative to other power supply sources, despite Group of Seven (G-7)[6] efforts to reduce use of older, unsafe Soviet-designed reactors. No one could conclude from these data alone that energy reform has worked.

Progress in reform is sometimes equated with progress in administering economic "shock therapy." The use of this term, borrowed from the old psychiatric practice of delivering high voltage to the temples of schizophrenics and thus rendering them stupefied if not catatonic for an indefinite period of time, is unfortunate.[7] In market reform, shock therapy has come to represent the only slightly more pleasant experience of rapidly imposed price, privatization, and budget reforms.

Perverse incentives created massive inefficiencies everywhere in the Soviet system and helped lead to its collapse. Anders Aslunde, the Swedish economist, has described the underlying economic stress on the Soviet economy that helped justify the 1991 coup attempt, which precipitated the change of state.[8] Some observers might attribute Soviet difficulties in the 1980s to the challenge of President Reagan's ballistic missile defense initiative, but the Soviet military could have matched U.S. spending on ballistic missile defense by increasing the central budget deficit by a mere 0.3 percent of estimated GDP. In comparison, the budget deficit of the Soviet Union under its last president, Mikhail Gorbachev, increased from less than 3 percent of estimated GDP in 1985 to more than 30 percent by 1990.[9] The 1998 budgetary crisis that led to Russian default and devaluation of the ruble was caused by a deficit one-quarter of that size. Market economies could not long have maintained Gorbachev's imbalances.

The Soviet deficit spiraled out of control when Gorbachev sought to provide workers with incentives to increase their productivity. He made the mistake of placing authority for increasing wages in the hands of managers whose every incentive was to increase pay. Salary increases were thus decoupled from productivity improvements. Although productivity and wages are linked by fundamental feedbacks in a market economy, productivity in the Soviet system had required either the idealism of its workers or the hand of terror. When the state lost the former

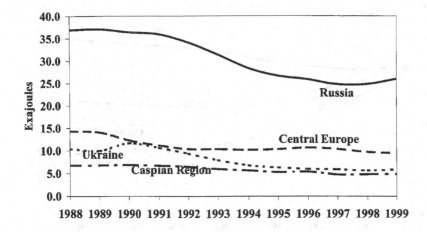

Figure 1.1   Energy Use in the Transition Economies, 1988–1999.
NOTE: Central Europe includes Bulgaria, the Czech Republic, Hungary, Poland, Romania, and Slovakia; Caspian Region includes Azerbaijan, Kazakhstan, Turkemenistan, and Uzbekistan.
SOURCES: BPAmoco, *Statistical Review of World Energy, 2000.* http://www.bpamoco.com.

and could no longer invoke the latter, declining productivity and stagnation resulted. General dissatisfaction with Soviet life also grew as the populace caught glimpses of higher living standards in the west. The periphery of the Soviet Empire started to break off when Poland's Solidarity movement challenged its authority, and the marshal law imposed in the early 1980s was too soft to control the rebellion. The Soviet world ended not with a ballistic missile but with a whimper.

In the early 1990s, economic collapse was the overwhelming reality throughout the region (see Figure 1.2). Russia's economic output fell by approximately one-third, depending on how barter trade and the underground economy are valued.[10] Soviet breakaway states experienced economic free fall in 1993 and 1994 when industrial output plummeted by one-quarter. Ukraine suffered even more. While decline continued at a slower pace in 1995, and by still smaller amounts in 1996 and 1997, recovery faltered. Russia had accomplished extensive privatization, but without imposing market accountability on the managers who wielded monopoly power and bought political clout. The government had tamed inflation but was unable to balance the budget. It continued to subsidize unprofitable industries, failed to collect taxes, and blundered by allowing the nonpayments crisis and barter trade to continue. The Asian financial crisis, plummeting oil prices, and a government budget deficit in 1998 in excess of 8 percent converged to crush the Russian stock market, obliter-

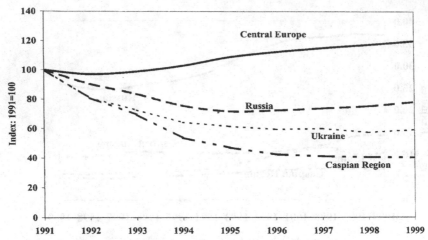

Figure 1.2   GDP per Capita in the Transition Economies, 1991–1999.
SOURCE: International Monetary Fund, *World Economic Outlook* (Washington, D.C.: IMF, 2000).

ate the ruble, and wither a budding economic recovery. Russia had failed, as much as anything else, to impose the discipline necessary to effect a market economy, the institutions, laws, and regulations the western world takes for granted. Russia may have been as close to a "free" market as any nation can come.

Fortunately for the credibility of economics, reform has seemed to matter. Nations implementing reform measures have done measurably better than those who have not implemented such measures. The World Bank has rated countries by extent of reform and characterized their progress in a kind of report card. At the head of the class were Poland, the Czech Republic, and Hungary. Belarus was clearly a D student, and the rest were in the C to C⁻ range. The A category freed prices, privatized more than half of production, and imposed hard budget constraints.

The "A students" of reform, the Czech Republic, Hungary, and Poland, did less poorly throughout the 1990s. Painful measures have helped Poland, for example, recover most of the economic ground it lost after the collapse of the Warsaw Pact trading bloc and to achieve high rates of economic growth several years in a row. Unemployment, however, was painfully high, as high as or even higher than in France. The A countries' economies never dropped more than one-quarter, although even that put them in a slump deeper than America's Great Depression of the 1930s.

These nations had resilience stemming from stronger cultural ties to western economies and closer proximity to Western European markets. Although some observers have argued that it was easier to reform a small nation like the Czech Republic or Estonia, population size did not predict economic performance. Hungary, the other Baltics, and Georgia have fared far better than other small nations, including Bulgaria, Romania, and Slovakia. Poland in 1998 had largely recovered, whereas somewhat larger Ukraine was still headed down. Measured in purchasing power parity, Poland's GDP per capita was over $6,000 in 1998, roughly twice that of neighboring Ukraine.[11]

The smaller nations did enjoy some advantages. Long before the changes, Hungary had relied on foreign trade for almost half its markets and had a market-oriented agricultural sector. Even if the land was not under the legal control of its farmers, they did control the inputs, however, and profited from good management and risk taking. Poland had a large contingent of entrepreneurs living in foreign lands and was relatively open. The Czech Republic had the advantages of relatively modern manufacturing facilities and a well-educated population. If the GDPs of the Caspian region[12] have fallen less in percentage terms, it is partly because they had less distance to fall. In purchasing power parity, Tajikistan, Turkmenistan, and Uzbekistan had fallen below $2,000 per capita by 1996. Although Armenia recently experienced a 5 percent turnaround, it was from a level of GDP of only about $2,200 per capita.[13] The economies and governments of Central Asia remain authoritarian and centrally controlled. Kazakhstan's 18 million people enjoy a GDP equal only to the budget of the U.S. Department of Energy. The Kazakhstanis put their hopes in the future of oil production.

The International Monetary Fund asserts that nations cannot long sustain budget deficits higher than 5 percent. Transition economies that reduced their budget deficits benefited economically within a few years' time, although the politicians who imposed austerity did not. A simple analysis suggests that the region's budget management—their deficits—may account for half the variation in later economic growth. Early fiscal prudence generally meant earlier recovery. Poland's 1992 deficit ballooned to 8 percent of GDP, but the nation quickly brought spending under control and since then has held its deficit under 3 percent. Both Poland and the Czech Republic have held their budget deficits below that of the average G-7 nation since 1992.[14] Similarly, the Czech government in 1997 faced a financial crisis[15] and responded quickly to reduce it, although recession followed and lingered through 1999. Hungary struggled to control deficits, which exceeded 5 percent of GDP in 1996–1997. Poland, the Czech Republic, and even Hungary outperformed the region in both deficit reduction and GDP recovery (see Table 1.3).

BOX 1.3   MEASURING ECONOMIC OUTPUT

Measuring energy intensity in an economy requires knowing its GDP. But GDP in transition economies is notoriously difficult to quantify, for at least three reasons. First, before the collapse of the Soviet Union, planners did not measure their economies in terms of GDP. They measured Net Material Product (NMP), an index consistent with materialist dialectics, but inconsistent with the notion of services and the service economy. Planners did not count services in the NMP, and so that index failed to count as much as one-third of "GDP." While GDPs are now routinely published for these countries, prior levels–the baselines against which change should be measured–can only be estimated. Second, black market and barter trade may account for as much as one-third of GDP in some countries, including Russia and Ukraine. Thus, both baseline and current economic estimates at best are imprecise and at worst are badly distorted. And third, GDP calculated at current exchange rates fluctuate with inflation and the value of the local currency relative to hard currencies. Such a measure has little meaning, at least in terms of real wealth.

Nations experiencing slow or declining GDP growth rapidly accumulated government debt.[16] Bulgaria, Ukraine, and Russia averaged negative growth in 1996 and 1997, and they had averaged budget deficits of roughly 7, 9, and 10 percent of GDP after 1992.[17] Belarus, otherwise the least liberalized country in the region, managed strong growth in 1998, although this was probably the result of monetary policy and is not sustainable.[18]

Budget deficits require large amounts of borrowing, and sovereign borrowers were considered relatively low risks, until Russia's 1998 financial debacle. However, government deficits draw funds away from private investment, particularly small business and its ability to create jobs, and government borrowing to cover the deficits elbows the private sector out of financial markets. Investors in 1996 ranked the Russian economy as riskiest for investment, even though it was the most profitable emerging market.[19] Its burden of debt meant high debt-service costs, which in turn meant insufficient funds to pay workers, including teachers and coal miners, and insufficient resources to solve other social problems. Economists well before the August 1998 financial crisis in Russia had warned that Russian tax reform had been inadequate to bring its budget deficit under control. That collapse further handicapped Russia and the former Soviet nations and increased the split between them and Central and Eastern Europe.[20]

Privatization has helped reduce inflation in most economies, surprisingly without making overall unemployment worse. But privatization has proceeded unevenly.[21] Poland, despite its favorable report card, has been slow to privatize state-owned enterprises. After an early wave of privatizations, the political process in Poland demanded a slower, more deliberate approach. Several countries tried voucher privatization, in which citizens are given the right to exchange vouchers for stock. Poland's answer was to delay the ultimate privatization until the capital markets developed. The Poles managed this in stages, beginning with commercialization, which is to say that firms were given hard budget constraints but their ownership was parked in state-controlled holding companies that managed them in a way that provided some capital and, perhaps more important, short-term survival.[22] These holding companies were ultimately to sell off assets to the public. The idea was to avoid the "fire-sale" consequences of depressed prices. This more deliberate approach was also more transparent, and both aspects helped to minimize corruption while providing some capital investment to ensure productivity improvements. Poland did, meanwhile, privatize over 2,300 state-owned enterprises.[23] Poland has done well economically despite applying a lower shock therapy. Confidence in the country has been indicated by quick sales of Eurobonds.[24] Hungary managed a much stronger privatization drive than Poland, but was driven to do so by a large budget deficit and the need to increase revenues.

The peripheral Eastern European countries Bulgaria and Romania at first followed a path more like that of the Soviet successors and experienced similar hardship. Neither the bloody revolution in Bucharest nor the bloodless one in Sofia countenanced much initial reform. On every index of liberalization—privatization, price decontrol, decentralization—southeastern Europe performed poorly, earning marks more like those of Ukraine than Poland.[25] These southern European nations were less open to western markets and less accessible to the outside world by air transport. Strong man Constantin Ceaucescu had distorted Romania's economy to a degree surprising even by Stalinist standards. Romanians sacrificed heat and power to save fuel for factories that produced exports and for Romania's "energy independence." Yet the energy system was so badly managed that the fires went out in Romania's power plants because the best coal was exported and available fuel was of extremely low quality. When a visitor in the mid-1980s asked what the heat rate in power plants burning low-grade fuel was, he was cautioned, "You realize you are asking for state secrets?"[26]

A hard currency crisis in 1996 precipitated a Romanian energy crisis when oil suppliers could not meet payments for oil imports.[27] Indeed,

Romania and Bulgaria have been described as "post-communism's biggest flops,"[28] although that ignominy may now go to Russia, where workers in underwear factories are sometimes paid in bras in lieu of wages.[29] Bulgaria did begin to recover after a 1997 election and the imposition of monetary and fiscal discipline. Belarus and Ukraine have been more deeply troubled. Officially, their economies were cut in half. Belarus, on 1 January 1998, raised the minimum wage to $7—per month[30] (half that of Russia).[31] Ukraine has admitted to official corruption and has suffered from domination by regional clans.[32] The U.S. government has withheld some Ukrainian assistance funding due to lack of reform progress in certain sectors, but certified that the nation had made progress with U.S. investors in resolving disputes over property ownership and the use of investment funds.[33] Ukraine, however, enjoys a profound advantage of peace, admirable given the potential for conflict with native Russians living in eastern Ukraine and with the Tatar ethnic minority in Crimea.[34]

East Germany's "instant transition" also did not go smoothly. The nation's labor and capital were highly uncompetitive but were suddenly forced to compete in the world's toughest markets. Only massive West German social transfers made the merger of the two countries work. The World Bank estimates that without the buyouts and early retirements made possible by the income transfers, unemployment would have reached 30 percent.[35]

One index of reform is confidence foreign investors have in nations. Poland and Hungary rank first and second in this measure, having attracted over $15 billion each in foreign direct investment between 1989 and 1998. The Czech Republic attracted about $10 billion over the same period, 10 percent more than much-larger Russia. Ukraine attracted less than one-fifth as much as comparably sized Poland.[36]

Reform and Energy Intensity

Some economists object that energy intensity is not a measure of economic efficiency, and technically they are right. High energy intensity could be economically efficient in places where energy costs are low, or when a less-developed nation must produce large quantities of energy-intensive basic materials, or when labor and capital costs are relatively high. Relatively low energy prices make energy less important than other inputs, and using more energy would be more economically efficient.

But the evidence is overwhelming that high intensity in the former Soviet Union pointed to economic inefficiency. Energy intensity in the United States in 1996 was about 13 megajoules per dollar of GDP, whereas Russia's was about three times higher. Russia's figure was distorted by the way GDP was measured and the way rubles were converted to dollars. However, compelling evidence of energy waste, in the economic sense, remains even after converting Russian GDP to purchasing power parity, after allowing for the comparative advantage of locally plentiful energy supply and for the structural differences between the west and the less developed nations of the former Soviet Union. Comparison of energy use per capita across countries provides more evidence of high energy intensity because usage is comparable to that in much richer countries (see Table 1.4). A large and growing literature has documented time and again that the planned economies failed to optimize energy, capital, and labor costs.[37]

The Soviet Union was far more material intensive than the market economies, but that fact reflected not a basic economic need, but the distortion of the economy away from higher-value goods, from a market point of view, to those preferred by the planners and the military.[38] As for energy prices, it is true that the Soviet Union should have enjoyed lower energy costs as a result of its abundance of gas and oil, but because energy was exportable to the rest of the world, world prices, minus transportation costs, might have prevailed under market conditions.[39] In other words, neither the notion that the Soviet Union's stage of development nor comparative advantage in energy resources explains the fact that energy intensity was two to three times that in the west.

The quality of energy itself also affects efficiency. Coal is especially difficult to burn effectively because it is bulky; has a low hydrogen to carbon ratio;[40] and contains considerable water, ash, and toxic compounds. Natural gas, on the other hand, has a very high hydrogen to carbon ratio, can be transported either as a gas or a liquid, and usually contains few impurities. Oil falls between these two fuels in quality and ease of use. Poland and the Czech Republic depended heavily on coal and were at a disadvantage. However, use of coal has declined by one-third to one-half, more than oil and gas, and this trend has improved fuel quality.

Energy intensity in the region was driven not so much by fuel type as by four economic factors stemming from central planning. First, industry dominated Stalinist economies, and heavy industry such as steel making requires more energy per unit of value added than light manufacturing or services, which have a higher ratio of labor cost.[41] Industry in the United States consumes less than one-third of all energy, compared to a sector average above 40 percent of end-use consumption in the region. This distortion of economic structure alone would make the formerly planned economies energy intensive, even if it were not for the antiquated state of industrial energy technology.[42]

Second, central planning grossly distorted price signals. Energy pricing under communism was just a way of keeping score. Prices had noth-

TABLE 1.4   Energy Use in the Transition Economies, 1999 (per Capita)

Country	Energy Use (gigajoules)	Electricity Use (kilowatt-hours)
Russia	176	5475
Ukraine	159	3915
Czech Republic	151	5900
Turkmenistan	146	3150
Hungary	104	3350
Poland	102	3480
Bulgaria	89	4950
Kazakhstan	86	3412
Belarus	99	2040
Uzbekistan	95	1785
Romania	64	2620
Azerbaijan	64	2080
Total/Average[a]	142	3900

NOTE: a) includes all transition economies, not just those listed. For comparison, consider that much richer Germany uses 143 gigajoules and 6206 kilowatt-hours per capita, respectively. Electricity data are for 1988.

SOURCES: Derived from British Petroleum, *Statistical Review of World Energy, 1999,* and United Nations Population Division, *World Population Prospects 1950–2050* (New York: United Nations, 1996); U.S. Energy Information Administration, www.EIA.DOE.gov/EMEU.

ing to do with cost and were rarely adjusted. To their credit, most transition economies have endured great pain to impose more realistic energy prices (see Figure 1.3). Although people's lives were rarely put in danger, energy importing nations have sometimes suffered from heat and electricity shortages.

Third, soft budget constraints, as described by Hungarian economist János Kornai, help explain the inefficient economic behavior induced by central planning.[43] In plain language, soft budget constraints meant that production cost did not matter, or, perversely, the-higher-the-better. Prices bore little relation to value and were little more than accounting terms. Enterprise budgets for labor, capital, and energy were allocated based on planned quotas, and bankruptcy for inefficiency and waste was nearly impossible. What market economists understood as the opportunity cost of capital, labor, and resources was not a concern for enterprise managers, whose opportunity cost was in not consuming planned allocations because that meant less allocation next year and less "profit" over-

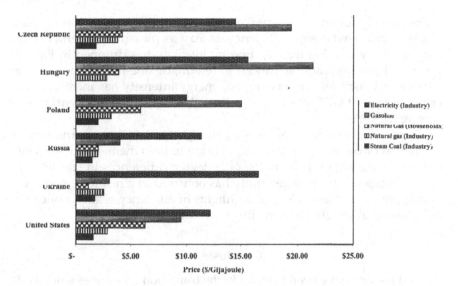

Figure 1.3   Energy Prices in Selected Nations, 1998.
SOURCE: *Energy Prices & Taxes–Quarterly Statistics* (Fourth Quarter, 1998), Part III, Section B, Table 12 (Non-OECD Countries) (Paris: International Energy Agency, 1999).

all. Benefits to the workers and managers—profit, in a sense—was calculated as a percentage of total costs. Josef Bognar, also a Hungarian economist, called this system "cost-plus pricing." If enterprise managers could negotiate a budget with the central planners in which they used twice as much steel and twice as much energy as was physically required, they would get more "profit" for bonuses, internal projects, or expanding their little empires. American utility managers might recognize this system, because regulators have set allowable profits for utility companies on the basis of cost-plus pricing. A power company earning 10 percent on the cost of every kilowatt hour sold naturally preferred to make power costing 2 cents to generate rather than 1 cent because its profit was twice as large.

Finally, technological development was stifled. Although the Soviet system trained and educated more scientists and engineers than the west, and these experts developed advanced technologies, the market did not demand and develop them. Price incentives and hard budget constraints in the west drive technological development as a way of cutting costs. Competition among suppliers drives technology for improving quality and adding new services. Incentives and competition, of course, have only recently been introduced in the transition economies.

Fortunately for environmental energy economists, reform has resulted in some improved energy efficiency, at least as measured in terms of energy intensity (see Figure 1.4). Energy intensity has dropped in Eastern Europe, the region most advanced in reform, by one-fifth. Conversely, in nations in which reform has lagged, energy intensity has increased. Energy per unit of GDP went up significantly in Ukraine and Central Asia during the 1990s.

Energy efficiency has not been a priority of either the governments of the region or outside governments seeking to help them. Progress in key nations in the region reflects to some degree differences in specific efficiency policies, but more generally has occurred as a result of macroeconomic policy. A brief review of highlights of efficiency-related policy developments in selected nations illustrates such differences.

Conclusions

Painful lessons have been learned in the transition economies since 1989. Energy use serves as an index of reform, as an indicator of successful and unsuccessful policies. The experience of Central Europe demonstrates that genuine reform works. Implementing hard budget constraints, meaningful prices, institutional reform, and economic restructuring has paid off in economic recovery and environmental improvement. Failing to implement these measures elsewhere has robbed citizens of economic and social well-being.

These conclusions all beg a critical question: Is the region better off now than before the collapse of communism? Except for a handful of Central European transition countries, the regional economy remains distressingly below pre-collapse levels.[44] Almost a decade passed between the onset of depression and the resumption of region-wide economic growth. Just as the region seemed set to recover, collapse in the financial markets shattered much of the painstaking progress made in that time. Central Europe, nevertheless, proves that reform can effect sustainable recovery.

Energy carries the ability to do work, create wealth, move armies, warm cities, and link the futures of nations. Some observers have suggested that both central planning and capitalism are flawed and that it is time to search for some third way, for some system halfway between communism and capitalism.[45] Railing against this notion, Vaclav Klaus, the former Czech Prime Minister, insisted that "visions of the third way are based on incompatible combinations of different worlds."[46] Energy sciences applied in the transition economies—a shop for choosing a world—reveal only what is at stake.

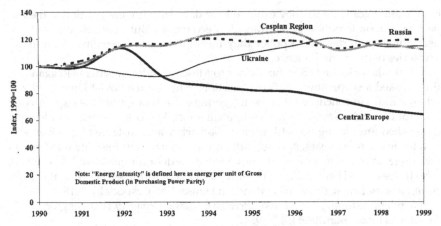

Figure 1.4   Energy Intensity in the Transition Economies, 1990–1999.

NOTE: Central Europe includes Bulgaria, the Czech Republic, Hungary, Poland, Romania, and Slovakia; Caspian Region includes Azerbaijan, Kazakhstan, Turkemenistan, and Uzbekistan.

SOURCE: Author, CO2 Emissions from Fuel Combustion, 1999 Edition (Paris: International Energy Agency, 1999), II.27; World Economic Outlook 2000 (Washington, D.C.: International Monetary Fund, 2000), 123; and BPAmoco, "Statistical Review of World Energy," www.bpamoco.com/worldenergy/.

Notes

1. Antonio Gramsci wrote this line during his incarceration by the fascists in Italy between 1926 and 1937. His notes, *Lettere dal Carcere*, published posthumously in 1947, were published in English as *Prison Notebooks*. Translated by Joseph A. Buttigieg and Antonio Callari (New York: Columbia University Press, 1992).

2. Francis Fukujama, "The End of History?" *National Interest* 16 (Summer 1989): 3.

3. See William Chandler, Alexei Makarov, and Zhou Dadi, "Energy for the Soviet Union, Eastern Europe, and China," *Scientific American* 262, no. 9 (September 1990): 121–127; see also William U. Chandler, *The Changing Role of the Market in the World's Economies* (Washington, D.C.: Worldwatch Institute, 1986).

4. Statement of Stuart Eizenstat, Under Secretary of State for Economic and Business Affairs, before the Senate Foreign Relations Committee, 22 July 1997.

5. Chandler, *Changing Role of the Market,* 17; Chandler, Makarov, and Dadi, "Energy for the Soviet Union," 121. In its simplest form, energy intensity is a measure of the rate of energy used to make physical products.

6. The G-7 nations include Canada, France, Germany, Italy, Japan, the United Kingdom, and the United States. The European Union participates in G-7 summits, and occasionally the group is expanded to the "Big-8" to include Russia.

7. The author, in a summer job in Tennessee during a break before becoming an undergraduate, helped administer shock therapy to mental patients. The violent, rapid convulsions they experienced with the electric shock and the passive be-

havior and vacant stares they exhibited for days after it were not unlike the impact of economic policy in the transition economies. But why economists have used the term to describe a policy they meant to advocate remains one of those mysteries of the dismal science.

8. Jack Matlock, the U.S. Ambassador to Moscow under presidents Reagan and Bush, asked a surprisingly difficult question: Why did the Soviet Union collapse when it did? He concluded that the triggering event was simply the hapless coup attempt by hardliners in 1991. A competent effort, Matlock asserted, would have dispatched foundering Soviet President Gorbachev and sustained the Soviet system for many more years. No coup attempt would have left the state intact for an indefinite time. The incompetent coup opened the door for President Boris Yeltsin, who for years had been positioning himself to lead reform. See Jack F. Matlock, Jr., *Autopsy on an Empire* (New York: Random House, 1995), especially 7–18.

9. Anders Aslunde, *How Russia Became a Market Economy* (Washington, D.C.: Brookings Institution, 1995), 47–49.

10. Official GDP statistics omit an unknown but large share of economic activity because barter trade and trade hidden from tax collectors do not get counted.

11. International Monetary Fund, *World Economic Outlook* (Washington, D.C.: International Monetary Fund, 1998), 133–135. Similar numbers have been published by the Austrian economics institute WIFO Volkswirtschaftliche Datenbank, http://www.wifo.ac.at/cgi-bin. See also *Business Central Europe*, a publication affiliated with *The Economist*, at www.bcemag.com.

12. The Caspian region here is defined as the former Soviet Republics of the Caspian region, including Armenia, Azerbaijan, Georgia, Kazakhstan, Kirghistan, Moldova, Tajikistan, Turkmenistan, and Uzbekistan. Traditionally, some of these would be considered Trans-Caucasus states. This book concentrates on the key energy states in this region: Azerbaijan, Kazakhstan, and Turkmenistan.

13. Interfax News Agency, *Trade & Investment Report* (7–14 April 1995); U.S. Department of Energy Information Administration, at http://www.eia.doe.gov/emeu/iea/append.html.

14. International Monetary Fund, *Progress with Fiscal Reform in Countries in Transition* (Washington, D.C.: International Monetary Fund, 1998).

15. Vincent Boland, "Big Losses Bring Down Czech Bank," *Financial Times*, 9 August 1996, p. 2.

16. Economists, of course, measure economic value with difficulty even in the best of times, and quantifying the output of economies in transition is fraught with danger. Pre-revolutionary output was exaggerated by the incentive to meet quotas. Post-revolutionary output has been understated by the incentive to avoid taxes. But few economists doubt the reality of deep decline in Belarus and Ukraine.

17. International Monetary Fund, *World Economic Outlook* (Washington, D.C.: International Monetary Fund, 1998), 133–218.

18. Belarus's growth in 1997 appears to be an anomaly. Perhaps the least-reformed nation, Belarus managed a 10 percent growth rate on the basis of monetary policy and regaining access to Russian markets, which helped the country restart large, inefficient enterprises. *RFE/RL Newsline* 2, no. 166, pt. II (28 August 1998); http://www.rferl.org/newsline/1998/08/280898.html.

19. "Russian Stock Market Best Among Emerging Markets in 1996," *Interfax Weekly Investment Report* (7 February 1997): 1.

20. European Bank for Reconstruction and Development, *Transition Report Update* (London: European Bank for Reconstruction, April 1999). See also Stefan Wagstyl, "Outlook Gloomy for Former East Bloc," *Financial Times*, 16 April 1999, p. 2; and John Thornhill, "Life Returns to Russian Economy's Corpse," *Financial Times*, 8 July 1999, p. 2.

21. Anonymous, "Lessons of Transition," *The Economist*, 29 June 1996, p. 81.

22. Other countries of the region also "commercialized" their state-owned enterprises. The key difference between Poland and the other Central European nations and Russia and Ukraine was the imposition of effective budget constraints. That is, enterprises had to balance their books without massive subsidies through budget transfers or nonpayments of utilities, taxes, and other obligations.

23. That is, Poland was first in the round of reforms in the late 1980s. Hungary and the Czech Republics had attempted reforms in the late 1950s and 1960s, respectively, and Poland had also made changes in the 1960s.

24. Richard Lapper, "Investors Snap Up DM205M Polish Eurobond," *Financial Times*, 10 July 1996, p. 13.

25. See generally World Bank, *From Plan to Market* (Washington, D.C.: World Bank, 1996).

26. Silviu Brucan, personal communication, Bucharest, April 1986. Brucan was a well-known sociologist who, in the late 1980s and early 1990s, played an important role as a dissident.

27. Virginia Marsh, "Romania Forced into Forex Controls," *Financial Times*, 13 August 1996, p. 2.

28. Anonymous, "Romania and Bulgaria: Those South-Eastern Laggards," *The Economist*, 19 October 1996, p. 54.

29. Penny Movant, "Far Eastern Factory Workers Paid in Bras," *OMRI Daily Digest* I, no. 14 (21 January 1997); http://www.rferl.org/newsline/1997/01/1-rus/rus-210197.html.

30. Anonymous, ". . . as Minimum Wages Fixed by Gov't at USD 7," *New Europe*, 25–31 January 1998, p. 4.

31. Penny Morvant, "Yeltsin Approves Minimum Wage Increase," *OMRI Daily Digest* I, no. 11 (16 January 1997); http://www.rferl.org/newsline/1997/01/1-rus/rus-160197.html.

32. Matthew Kaminski, "Ukraine Held Hostage by the Clash of Its Clans," *Financial Times*, 7 August 1996, p. 2; John Thornbill, "Kuchma 'Sacks' Premier," *Financial Times*, 20 June 1997, p. 2; Ustina Markus, "World Bank Protests Corruption in Ukrainian Government," *OMRI Daily Digest* (9 January 1997); http://www.rferl.org/newsline/1997/01/3-CEE/cee-09197.html; Strobe Talbot, "Countering a Communist Comeback," *The Washington Post*, opinion, 13 April 1998, p. A23.

33. Anonymous, "Albright Certifies Ukraine Progress on Investment Complaints," State Department press release, 29 April 1998.

34. Chrystia Freeland, "Eastern Ukraine Turns Its Back on Russia," *Financial Times*, 23 July 1997, p. 2.

35. "Developing the Energy Sector in the Transition Economies of Central and Eastern Europe and the Former Soviet Union," World Bank, http://www.worldbank.org/html/extdr/offrep/eca/ecaeng.htm and http://www.worldbank.org/data/countrydata/pol-aag.pdf.

36. European Bank for Reconstruction and Development, cited in Stefan Wagstyl, "Unafraid to Change," *Financial Times*, 15 September 1999, p. 15.

37. See Chandler, Makarov, and Zhou, "Energy for the Soviet Union, Eastern Europe, and China," 121–123; Dale Gray, "Reforming the Energy Sector in Transition Economies: Selected Experience and Lessons," *Europe and Central Asia Region* World Bank, World Bank Discussion Paper (Washington, D.C.: World Bank, 7 July 1995); Igor Bashmakov, *Energy for the New Millennium* (Moscow: The Center for Energy Efficiency, 1999), 26–28; See generally World Bank, *From Plan to Market: World Development Report 1996* (New York: Oxford University Press, 1996).

38. Note also that under the quota system, it was advantageous to exaggerate economic output. However, correcting for that fact would only increase the relative intensity of the east. Services were not even measured as part of the "Net Material Product" and had to be estimated to compare the total economic output—the GDP—of the planned economies with nations of the west. Estimating Purchasing Power Parity was one way to do that. Russian economists have considered these factors and found that no matter the stage of the economy, value of output, or cost of energy, the planned economies did in fact rank as the most energy intensive in the world. This result was not surprising to market economists who believed that markets most efficiently set prices and that prices most effectively delivered information about the value of goods and services, including energy and heavy materials. Prices, of course, were virtually meaningless in the planned economies, and energy and materials were allocated by the quota system. Often, if one economized and became more efficient, one was penalized with a lower allotment in the subsequent allocation.

39. An argument can be made, however, for lower domestic oil prices in Russia as a function of limited export capacity.

40. It is the hydrogen in fossil fuels that provides energy during combustion. Among coal, oil, and gas, gas produces the highest amount of energy per carbon atom because it has the highest hydrogen to carbon ratio (4 to 1 for methane, or CH_4, which is the principal constituent of natural gas).

41. William Chandler, *Changing Role of the Market*, 16–21.

42. See generally World Bank, *From Plan to Market*.

43. János Kornai, *Contradictions and Dilemmas* (Cambridge, Mass.: MIT Press, 1986), 8–9 and 26–27; see also Chandler, *Changing Role of the Market*, 17.

44. Stefan Wagstyl and Charles Clover, "E Europe Pressed on Financial Controls," *Financial Times*, 9/10 May 1998, p. 3.

45. For an indicator of interest in this notion, see Jean-Claude Casanova, "Les Socialistes et la 'Troisieme Voie' [Socialists and the 'Third Way']," *Le Figaro*, Paris, 14 October 1998, p. 2.

46. Vaclav Klaus, "Interplay of Political and Economic Reform Measures in the Transformation of Postcommunist Countries" (presentation to the Heritage Foundation, Washington, D.C., 15 October 1993).

2

End-Use Efficiency

The planned economies were, in principle, engineered for energy efficiency. Energy did not provide lavish creature comforts. Soviet apartments always offered less than half the living area of U.S. homes.[1] Planners built housing in multifamily apartments that should have been more energy-sparing than U.S.-style housing.[2] Moreover, planners engineered heating systems that served many buildings from a single central heating boiler. However, energy inefficiency—in the economic sense of the term—existed throughout the history of the Soviet Union and prevailed everywhere in the region. Russian economist Igor Bashmakov once described Russia as "the Saudi Arabia of energy efficiency."[3] Soviet steel mills in the 1980s, for example, used 1.5 tons of coal to make a ton of steel; Japanese mills used half as much. Similarly, Polish cement required twice as much energy per ton as French cement. The region's inability to get more service and more production stemmed from policy but was evident in the technology of buildings, industry, and transportation.

Buildings

Poland's demand profile in the buildings sector is fairly typical for the region. Energy use in residential and commercial buildings in the late 1990s totaled just under one-fifth of national energy demand, a much lower share than in the west (see Table 2.1). Up to four-fifths of this energy was used for basic needs such as heat and hot water. Of the remainder, lighting and cooking required about 10 percent each. Polish homes used—and use—twice as much energy for a given amount of area as Western European ones.[4] Changes in energy use in the buildings sector should reflect a shift to a more consumer-oriented society, with household energy services as well as commercial energy services, particularly retail operations, included in the total. The buildings sector's share of final energy use in Poland, however, declined slightly in the 1990s, suggesting either

TABLE 2.1   Energy Use by Sector, 1985 and 1995 (percent)

Country	Industry		Transportation		Buildings[a]	
	1985	1995	1985	1995	1985	1995
Belarus	34	35	15	15	33	33
Bulgaria	51	59	10	5	14	28
Czech Republic	56	48	8	14	21	25
Hungary	39	24	11	16	37	52
Poland	43	39	9	14	38	34
Romania	58	57	4	13	14	22
Russia[b]	56	38	15	11	17	31
Ukraine	50	48	9	8	33	38

NOTES: a) Combines public sector, residential, and commercial energy use. Totals do not add to 100. Agriculture is less then 8 percent. b) Data for 1985 are for the entire Soviet Union.

SOURCE: International Energy Agency, *Energy Balances of Non-OECD Countries, 1996–1997* (Paris: Organisation for Economic Cooperation and Development, September 1999); International Energy Agency, *Energy Balances of OECD Countries, 1996–1997* (Paris: Organisation for Economic Cooperation and Development, June, 1999).

that restructuring was unnecessary or that increases in value added have come as a function of productivity improvements.

District systems provide 75 and 65 percent of space heat in Russia and Ukraine, respectively.[5] Apartments receive heat from a central heating station via heated water pumped through large diameter pipelines sometimes tens of kilometers long. The large boilers in the central plants were designed to provide economies of scale over individual heating units. It is ironic that Soviet-style housing used—and uses—two to three times as much energy to heat a given amount of floor space as U.S. single-family, individually heated homes. Some experts estimate the heat loss in the pipelines at 40 percent of total heat production, but this is highly uncertain. The severity of the problem in Moscow has been illustrated by people falling to agonizing deaths in scalding sinkholes created by leaking hot water pipelines.[6]

Buildings throughout the region lacked basic energy efficiency measures. Although apartments were small, they were built of materials such as cement blocks or prefabricated concrete panels that readily conducted heat to the outdoors. Recent tests on apartments built in Russia during the reign of Soviet Premier Khrushchev reveal an "R-value" of 1 in the walls. The R-value required in Czechoslovakia for new buildings in 1979 was 1.8 for roofs and 0.95 for walls.[7] This R-value is a measure of thermal integrity; the higher the number, the more insulating the material. A U.S. home built in 1980 in a cold region—Minneapolis, for example—was required to be built with wall insulation equal to R-11. Moreover, even then the cost-effective value in such a cold climate with U.S. prices was probably closer to R-19.[8] In the Khrushchev-era apartments, heat escapes on the cold northern exposures of buildings and accumulates on the southern exposures, creating discomfort on both sides of the buildings.

Another factor contributing to heat energy losses is lack of basic sensors and controls on heating plants. Plant operators across the region still decide how much heat to supply on the basis of the weather forecast. A western home often has a thermostat set to turn a furnace, boiler, or heat pump on and off depending on demand, as indicated by the temperature reading at the thermostat. Few such readings are available to the central boiler houses of the post-planned economies. Plants were built without meters or sensors indicating demand for heat. Demand could be only inferred from the temperature of the return water flowing into the boiler house from the pipes and buildings. Even if customers were too hot and wanted to control the heat, they lacked thermostats to do so. Everyone who has been in Soviet-style housing knows the joke about the *fortochka*, the little window serving as the thermostat—and that it is not a joke.

Household radiators often are much hotter—and much colder—than needed. This situation sometimes arises because the last flat served by the pipe needs to be as warm as the first one, yet each unit served along the way will reduce the available heat somewhat. This problem can be remedied by the use of heat exchangers, located at heating points where large diameter pipes shift heat to smaller diameter pipes. Separate pipes can go to separate buildings, instead of one pipe serving many buildings. When heat exchangers are present, they often are tube-in-tube arrangements that simply consist of one relatively thick pipe inside another. At the entrance to each building, another heat exchanger can be used to step down the temperature to that required at any given moment. Heat exchangers remain inefficient, at least compared to modern fin-type heat exchangers, devices that resemble car radiators. Heat transfer is often impeded further by accumulation of minerals and debris in the tubes. These accumulations become so severe that pumps on the district heating system use most of their power in overcoming the resistance of accumulated solids to the flow of water.

Appliances differ significantly in size and performance from European and, especially, U.S. ones. For example, a typical Polish refrigerator-freezer in the early 1990s provided about 300 liters of volume, or 10.5 cubic feet. A typical U.S. refrigerator-freezer is twice that size. The Polish model consumed about 750 kilowatt-hours per year, about the same as a new U.S. machine of twice the volume.[9] However, before appliance efficiency standards were enacted in the United States in the 1970s, the typical U.S. refrigerator-freezer consumed about 1,500 kilowatt-hours per year. Most Polish households have refrigerators, and volumes will likely increase as incomes increase. However, Poles are likely to take advantage of newer technologies to reduce the per unit volume energy costs to the better U.S. levels. Moreover, harmonization with European "standards" will likely mean a convergence to European efficiency levels.

Industry

The Soviet energy system mainly served industrial and military production. Industry was structured for a heavy industry base: metallurgy, defense materiel, basic chemicals. The largest energy-consuming industrial sectors were iron and steel, chemicals, cement, and aluminum. These primary industries required large amounts of energy, particularly compared to their value, making them by definition "energy intensive." Light manufacturing was never a priority, and structural reform has been slow in coming. In recent years, industry has represented more than half the economic output of the planned economies, whereas European and U.S. industry has represented only about one-third. The service sector in the United States generates about half of the GDP, but in Russia it produces less than one-fifth.[10] One indicator of the level of services and the energy services required to provide them is commercial floor space per capita. In the Soviet Union, this indicator totaled about 5 square meters, compared to 10–15 square meters in Europe and 25 square meters in the United States.[11]

Industry's share of energy use in Hungary fell after economic reforms of the 1990s from about 40 to about 25 percent of total final energy use (see Table 2.1). Poland's industrial energy share had fallen slightly by 1995 but still remained at almost 40 percent. That figure suggests that Poland's energy intensity reduction did not come about as a result of a structural shift—in the sense of shifting to services—but rather from increased productivity, including increased energy productivity. Anecdotal evidence from the steel sector supports this idea. The most antiquated mills were shut down while the more modern ones were supported. Unfortunately, the data may simply be too variable or unreliable to draw firm conclusions.[12]

Industry today remains technologically backward, and energy per unit of physical production has changed little since 1990 (see Figure 2.1). Soviet technologists did not utilize fundamental technological improvements for iron and steel production, as did market economies. A classic case of technical change involved the replacement of the open hearth furnace with the basic oxygen furnace. Both furnaces serve the same purpose, making metallic iron by reduction, or removal, of oxygen from iron ore. Both furnaces heat coke, made from petroleum or coal, together with ore in a reducing, or low-oxygen, atmosphere. Coke combustion produces carbon monoxide, which combines with the molecular oxygen in iron ore—ferric oxide—to produce molten metal. In the open hearth furnace, the oxygen level was controlled with regular air. In the basic oxygen furnace, pure oxygen was blown into the furnace in the ratio needed for more efficient production. The western steel producers retired most of their inefficient open hearth furnaces decades ago in favor of the more efficient basic oxygen approach. The oxygen furnace was first introduced in the west in the 1950s and 1960s to reduce energy costs and increase productivity. Although the open hearth furnace had been largely phased-out in the west before the Soviet Union collapsed, it still accounted for roughly half of production in major transition economies. In Poland, however, steel production using the open hearth furnace fell from 39 percent in 1988 to only 11 percent in 1996.

A new wave of innovation in the steel industry followed the introduction of the electric arc furnace. The technology relies primarily on recycled steel instead of virgin iron ore, using electricity to melt the scrap. Recycling steel by melting it in the electric arc furnace requires only half as much energy as producing steel by reducing virgin ore with coke. Western nations have widely adopted this technology to reduce costs and increase profits. The renaissance of the U.S. steel industry in the 1980s was to no small extent due to the rapid rise of the "mini-mill," which married the electric arc furnace with new techniques of metal rolling to produce higher-quality products. The firm Nucor, in North Carolina, exemplified the profitable modern steel company, primarily by using mini-mill technology. The United States and Japan used this "recycling" furnace to produce over one-third of output, whereas the Soviet Union in its last years used it for only one-ninth of production.[13] To the extent this ratio has changed, it is only because the most inefficient plants have been closed, not because new technology has been installed.

Another aspect of restructuring will affect the metallurgical sector: improving the strength and quality of metals so that manufactured products can be lighter. One study estimated that in 1990 Soviet manufactured goods were 10–70 percent heavier than western products because Soviet materials manufacturing methods were less advanced.[14]

The Soviet Union failed to innovate advanced technology even when Soviet scientists invented the technology. A classic illustration of this failure is continuous casting of steel. This practice takes molten steel directly from the furnace and casts it into desired forms before the steel cools. If

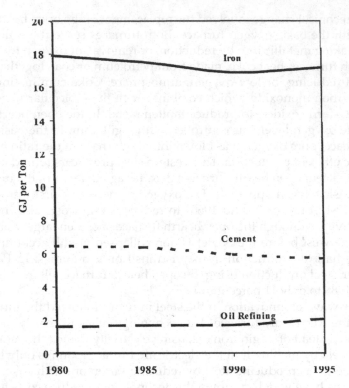

Figure 2.1   Energy Used per Ton of Product Produced: Russia, 1980–1996.

SOURCE: *Russian Energy Picture* (Moscow: Center for Energy Efficiency, 1998).

metal is allowed to cool and solidify before being cast, it has to be re-heated and then cast. Continuous casting, invented in Russia decades ago, accounted for only about 17 percent of Soviet steel made from virgin ore, compared to 90 percent in Japan.[15]

Western chemical industries dramatically improved energy productivity during the 1980s and 1990s. The innovations behind that change came about partly due to the pressure of oil price increases of the 1970s—oil and gas are the primary feedstock for chemicals—and partly due to the need to be very efficient to compete with growing production in low-cost areas, particularly the Middle East. The Soviet sphere was largely insulated from both these pressures, however, and did not apply techniques widely adopted in the west, including advanced membranes, catalysts, or even energy management practices such as heat recovery or insulation. As with metallurgy, the only technical change has been to shut down the most inefficient, outmoded systems.

Soviet technical innovation also lagged dramatically in the oil refining sector. Distillation of fuels from crude oil takes advantage of the different boiling points of different molecules in crude oil. When crude oil is heated in a distillation column, light compounds, including propane, boil off first, followed by gasoline, diesel, and kerosene. By augmenting distillation and adding cracking or catalytic processing of crude oil, more useful products can be produced. However, catalytic cracking in Russia still accounts for only a small percent of refinery output.[16] Indeed, Soviet era refineries still produce more *mazut*—what Americans call residual fuel, resid, or No. 6 heating oil—than any other product. Residual oil is useful mainly for power generation and for making asphalt. In the west, residual oil amounts to 5–10 percent of output, but Soviet-designed refineries end up with 40–50 percent of output as resid. Cracking units—facilities that use catalysts to break up and reform heavy molecules into more useful lighter products like diesel—have rarely been installed. Advanced distillation techniques have also been neglected.

Construction materials and cement production provide another dramatic contrast with western practice. French firms have moved almost exclusively to the production of cement using the "dry process," which requires less water evaporation when the product is calcined, or heated to form cement from calcium carbonate. It uses 20–30 percent less energy than the "wet" process. The wet process still accounted for 85 percent of production in the transition economies in the early 1990s.[17]

Industrial energy management in the transition economies has lagged behind western practice across all sectors. For example, fuel and air should be mixed properly for efficient combustion. Insufficient air results in incomplete combustion, and excess air carries heat uselessly up the smoke stack. Setting this ratio properly requires modern equipment to sense the molecular ratios and to adjust valves to provide a "stoichiometric" combination. But eastern boiler house operators lack such devices and must improvise, often using visual cues to adjust their settings. It is common to see operators peering through a small window in the furnace wall to observe flame color and shape and adjusting the fuel and air flows accordingly.

Similarly, steam systems should be equipped with functioning steam traps. The steam trap has been described as a perverse but essential device that prevents steam from escaping from steam pipes while allowing condensate—and dirt—to pass on to return to the boiler. Steam lines east of Budapest have for decades lacked working and well-maintained steam traps. Retrofitting steam systems now provides rapid pay back, on the order of a few months. But financing for all such fixes is limited by economic transition, an environment more hostile than any reduction furnace and more constraining than any trap.

Transportation

Citizens of the former Soviet Union enjoyed only one car for every 17 people, compared to one for every 2 Germans and 1.7 Americans.[18] A civil defense era joke had it that if the United States had to evacuate its cities, everyone could get in a car at the same moment and no one would have to sit in the back seat.[19]

Before the end of the Cold War, mass transit was the dominant personal transportation mode in the planned economies, and passenger cars were very limited in number. They also rated low in fuel economy. The Russian-made Lada or Polish-made Polonez consumed 8–10 liters of gasoline per 100 km (20–27 mpg), in contrast to a comparably sized Honda Civic that used 5–6 liters per 100 km (35–45 mpg) or a Ford Escort that used 6–7 liters. The difference, more than anything else, was technology: four-valve cylinders, special cams, fuel injection, and computer-control technology the west now takes for granted but that has yet to penetrate eastern markets, except via imported cars.

Usage patterns in the region differed most dramatically from the west and are changing most rapidly. The number of cars in the region has increased dramatically, mainly through the importation of used vehicles.[20] In the Czech Republic, the number of cars has increased by 40 percent since 1990, taking it to about two-thirds the level of the average in Western Europe. The number of cars in Eastern Europe and Russia will continue to rise, probably approaching the Western European level within a decade.[21] However, automobile use had already been increasing rapidly in the command economies for a couple of decades before the recent changes. In Central Europe, automobile ownership—not to say use—had increased between 10 and 20 percent per year in the years 1970–1990[22] (see Figure 2.2).

Along with electricity, gasoline is the most expensive common energy carrier in the world today. Gasoline in Central Europe costs twice as much as in the United States and costs more than electricity in either region. GDP in Central European nations ranges between $2,000 and $12,000 per capita, one-quarter to one-half that of the United States, and income elasticities hold regional energy use below that in the west. Automobile use in the Czech Republic grew in the 1990s at an annual rate of 6 percent, mainly at the expense of passenger mass transit, which shrank almost 10 percent per year. Polish passenger transit has decreased 7 percent annually since 1988. The Polish government, moreover, has committed to adding 6 percent to the total highway length, or about 2,000 kilo-

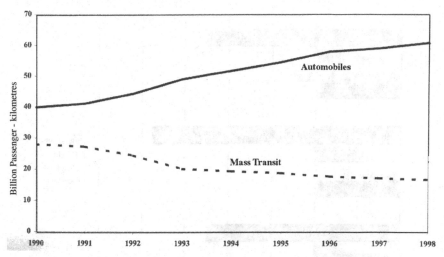

Figure 2.2   Automobiles Versus Mass Transit in the Czech Republic: 1990–1998.
SOURCE: Ivana Svotjkova, SEVEn, personal communication, Prague, Czech Republic, 23 November 1999, citing Czech Center of Transport Research, May 1999.

meters. The government hopes the private sector can be persuaded to provide equity financing for half of its planned highway expansion.[23]

The growth in the numbers of cars owned in Russia accelerated from over 7 percent annually in the 1980s to almost 9 percent in the 1990s.[24] This acceleration was most evident in the expensive imported luxury cars, but was also seen in Russian-made automobiles and imported used cars. The latter, however, have been heavily discouraged by high import duties.[25]

A similar shift from rail to road freight has occurred in Central Europe (see Figure 2.3). Significantly, this has not happened in Russia, by far the largest market. A key to future transportation energy use in the region could well be the management of the Russian railroads. The former Warsaw Pact region has fewer roadways than Italy.[26]

Mass transit stands out as a possible blessing of central planning, if well-maintained subways, trams, and buses can move people more quickly and safely in the densely populated cities of Eastern Europe and European Russia. Anecdotal evidence suggests a different reality, with poor maintenance and under-financing causing overcrowding and unsafe operation. Only heavy capital investment and proper financing will ensure survival of services that can compete with automobiles if, as is likely the case, even congested highways deliver passengers quicker and

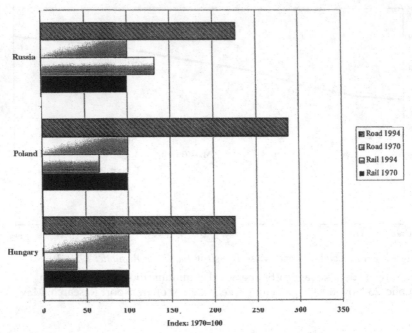

Figure 2.3   Freight Transport, Selected Countries, 1970 and 1994.

SOURCE: Organisation for Economic Cooperation and Development, *Transportation Statistics 1997 Annual Report* (Paris: OECD, 1997), 257–258.

more comfortably. On the other hand, massive state investment would be required for expansion of highway transportation. A public debate on transportation priorities will be a logical consequence of these needs in most democratic societies.

Technology Cost Analysis

The benefits and costs of efficiency technology have been evaluated in depth in Poland, a nation with demand-side needs and constraints typical of the post-communist countries. A study by the Polish Foundation for Energy Efficiency (FEWE) estimated the internal rate of return on sixteen technologies for energy efficiency in the industrial, power generation, and buildings sectors, and screened them for profitability.[27] The criterion for estimating the sum total of Poland's efficiency potential was set at an internal rate of return (IRR) of 18 percent. That is, the nation's potential for energy savings was tallied using only measures that were more profitable in real terms than Wall Street has been over the past five or six decades. The analysis was made using energy prices at their 1992 levels, and because energy prices have continued to rise, underestimates the value of the savings potential. The price of industrial and steam coal was assumed to be $35 per ton, and industrial and residential electricity was assumed to cost $0.04 and $0.049 per kilowatt-hour, respectively.

The "technical potential" in the 1990s totaled about 1 exajoule, or one-fifth of 1991 primary energy consumption. That figure, of course, is not particularly illuminating because it says little about the practicality of implementation of the technical options. FEWE estimates the profitable, achievable potential for efficiency in Poland to be about 10 percent of 1991 demand, or over 0.4 exajoules. This, of course, also begs many questions about the most serious obstacle to capturing the energy efficiency potential of the region: finance. The most profitable technologies identified by FEWE include:

- Steam traps: IRR[28] → 335 percent
- Polish gas residential boilers: IRR → 210 percent
- Low pressure sodium street lighting: IRR → 155 percent
- Industrial electric motors: IRR → 55 percent
- Heat meters and controls: IRR → 40 percent
- Compact fluorescent lights: IRR → 18 percent.
- Gas turbines for coal bed methane and biogas: IRR → 18 percent.

Technologies not deemed highly profitable with current prices but attractive with future prices (at Western European levels) include the technologies considered above, but not fluorescent lamps in multiple mountings, low-emissivity windows, integrated coal gasification power stations, diesel combined cycle generation, municipal solid waste power stations, or atmospheric fluidized bed coal combustion.

The value of profitable energy savings with current prices in the mid-1990s totaled over $2 billion per year in Poland alone. Much of this value, however, will go unrealized due to a host of institutional constraints. Despite the attractiveness of efficiency, few major investments have been made in delivering energy efficiency in Eastern Europe, the former Soviet Union, or China, for that matter. Opportunities within virtually every application—from steam traps to turbines—go begging.

Industrial energy efficiency in Ukraine is a priority.[29] Energy now totals 30–70 percent of total enterprise production costs, but few enterprises can obtain the capital necessary for large-scale energy efficiency improvements.[30]

Poland

Poland in the earliest years of reform underwent the same increase in energy intensity that Russia and Ukraine have recently experienced. The economic fall was too sudden for energy utilization to be adjusted commensurately. Shutting down the most inefficient (in an overall economic sense) producers required more time for adjustment. However, Poland stabilized after only three or four years of transition, much less time than it took Russia and Ukraine to stabilize. Energy intensity actually fell in the mid- to late-1990s and is expected to continue to do so if Poland maintains its commitment to reform.[31] Ukraine, meanwhile, has continued to experience energy shortages and difficulties importing fuel.[32]

Poland took a tough but deliberate approach to energy reform, imposing hard budget constraints and price rationalization early, but not through privatization and complete decontrol. The Polish government has since taken a big step toward the latter with its 1997 passage of the Energy Act.[33] The Ministry of Finance in 1999 gave authority over energy prices to a newly created Energy Regulatory Authority. The average prices of electric power and gas, respectively, remained 40 and 18 percent below replacement costs in 1997.[34] District heating utilities have been made self-financing, and less than 2 percent of supply costs remained subsidized in 1998. Indeed, heat energy prices increased to the point that heat cost 14 percent of average household income, several times the share of household income spent on heat in Western Europe.

Coal was priced close to the border price, but still slightly below production cost. Oil products, particularly motor fuels, became much cheaper than in Western Europe, but mainly as a function of tax differentials. Although it was less severe than in Russia and Ukraine, Poland developed a utility nonpayments problem, with up to one-quarter of electric and heat bills going uncollected.[35]

Poland attempted its first energy efficiency policy reforms in 1992 in the last days of the liberal Bielecki government. However, when the liberal government lost to a post-Communist one, reform policies—privatization, price decontrol, and measures to promote efficiency—were slowed considerably. The post-Communists, however, showed a surprising degree of flexibility, probably because the Polish electorate clearly wanted change. Although privatization of utilities and large, state-owned enterprises progressed slowly, the rest of the economy was generally freed to develop along market lines. Inflation was reduced, with moderately stringent monetary policy and reasonable fiscal deficits. Foreign trade was promoted, and competition from imports provided price moderation on the part of domestic producers. A legal and institutional infrastructure developed rapidly, and Poland thus became the first country of the region to show positive economic growth.

The new energy law phased out price controls over a two-year period. It also ensured access to the utility grid for independent power producers and liberalized the oil and gas sector. Poland had tightly controlled the petroleum sector by giving monopoly status to the Polish Oil and Gas Corporation,[36] but following the 1997 energy act, exploration and development of coal-bed methane, natural gas, and oil was permitted. Duties on imported fuels were reduced.[37]

Czech Republic

The Czech government offers a surprising combination of good and bad energy policy examples.[38] Prime Minister Vaclav Klaus developed a reputation for being more conservative than his western tutors. For example, Klaus shocked the World Bank by refusing to provide sovereign guarantees for World Bank project loans—loans that were virtually approved—totaling a quarter of a billion dollars. The irony of an Eastern European tossing the World Bank out of his country for being too interventionist was not lost on the business community, which for a time considered the Czech Republic as the most market-oriented economy in the region. The Czech government, nevertheless, slowed privatization of CEZ, the utility monopoly, approved a guarantee for the Temelin power plant, controlled heat prices to avoid political criticism just before elections, and was slow to divest its municipal heating utilities. The Czech Energy Efficiency Center, SEVEn, estimated that in 1997 subsidies for the supply side outweighed those for conservation by over ten to one (see Table 2.2).

However, the Czech Republic has implemented a series of measures intended to improve the efficiency of energy use. These measures include:[39]

- *Building thermal energy standards:* The central government in 1994 adopted a voluntary standard that updated and significantly strengthened a 1979 measure.[40] The standard is generally met or exceeded.
- *Heat and hot water metering*: Metering was made mandatory in a 1991 decree, which also mandated thermostatic radiator valves and cost allocators on all radiators. The valves and allocators requirement, however, was removed by a 1995 Ministry of Industry decree that required their installation only at the time of upgrading of the heating system.

TABLE 2.2   Subsidies: Supply Versus Demand in the Czech Republic, 1997

Subsidy	Million $ U.S.
Demand side	
Energy efficiency demonstrations	$8
Fuel-switching grants for clean air	$50
Total, demand side	$58
Supply side	
Residential district heat subsidies	$270
Cross-subsidy to residential power	$400
Cross-subsidy to residential gas	$100
Total, supply side	$770

NOTE: Czech heat subsidies were eliminated in 1998.

SOURCE: SEVEn, The Czech Energy Efficiency Center, Prague, 1998.

- *Tax incentives*: The general 22 percent value-added tax was reduced to 5 percent for selected energy-saving devices such as compact fluorescent bulbs and for renewable energy systems such as low-head hydro. The tax reduction for renewables, however, was eliminated in 1998.
- *Clean Air Act*: Performance standards in 1991 were imposed on boilers and other stationary power sources, and the government was given authority to close facilities for noncompliance.

The Czech government also innovated a model program of air pollution control in which over a quarter of a billion dollars was invested in efficiency and fuel switching—to gas—to reduce the use of dirty coal for heating and power units. Coal had long supplied the Czech Republic with three-quarters of its energy and was so filled with toxic materials that children living near the plants exhibited stunting and bone defects.

Against these success stories, however, CEZ, the electric power monopoly, now promotes electric power use and even encourages space heating using electricity. Unlike in Poland and Ukraine, access to the power grid for competing sources of electric power generation is not assured. The Czech government has been considering an Energy Efficiency Act since 1995, but has not yet acted on it.[41] All heat supply subsidies were eliminated in 1998, although prices may not be increased more than 3 percent per year without the permission of the Ministry of Finance.[42]

Russia

Russia's road to reform has suffered a landslide or two along the way. The government chose to protect consumers in the earliest phase of liberalization, and although energy prices increased, they did so more slowly than the overall price index.[43] Nevertheless, prices became meaningful for the first time, increasing by a factor of 5–7 for all energy carriers. As a share of income, energy in early 1998 cost more than it did in Europe or the United States.

A decentralized approach to efficiency policy in Russia initially offered much promise. Demonopolization and rationalization of the energy sector have been implemented at the federal level, with leadership for financing, standards, and public education taking place at the regional level.[44] This policy includes the creation of innovative efficiency funds in Chelyabinsk, Nizhni Novgorod, and the Komi Republic. Several dozen Russian local governments have implemented decrees and legislation for efficiency, although these often tend toward command and control measures without either the means or the technical basis for enforcement. More usefully, many local governments have also created energy efficiency offices and centers. These operations have mobilized government resources, including some financial backing, to place priority on efficiency improvements, particularly in the industrial sector. Tax incentives, however, rarely have been applied, and the confiscatory nature of the Russian federal and local tax system all but drove the economy underground, diminishing the prospects for financing. After the August 1998 financial debacle, the importance of reform and of improving the well-being of average Russians only increased.

Building codes and standards for consumer protection and cost-effectiveness are being implemented in Moscow, Rostov, and Yaroslavl, and are being developed in Chelyabinsk and Omsk (see Table 2.3). In the United States, local government is responsible for building codes. For example, states or cities set standards for the energy efficiency of almost all the buildings constructed. National and international agencies are working on developing such standards for the transition countries. CENEf, with the help of the Natural Resources Defense Council, a U.S. scientific and legal environmental advocacy group, has worked with the Russian cities to enact model local codes. Thermal standards for high-rise apartment buildings were implemented in Moscow beginning

TABLE 2.3   Energy Efficiency Policies in Russia's Regions, 1999

Region	Department for Efficiency?	Efficiency Loan Fund?	Building Codes?	Budget Line Item?	Tax Incentives?
Chelyabinsk	Yes	Yes	Yes	Yes	No
Moscow	Yes	Yes	Yes	Yes	Yes
Nizhni Novgorod	Yes	Yes	No	Yes	Yes
Tulskaya	Yes	No	Yes	Yes[a]	No
Tomsk	Yes	No	No	Yes	No
Yaroslavl	Yes	No	Yes	Yes	No
Samara	Yes	No	No	Yes	No
Kostroma	Yes	Yes	No	No	No
Murmansk	Yes	Yes[a]	No	No	No
Omsk	Yes	No	Yes[a]	No	No
Sverdlovsk	Yes	No	No	Yes	No
Kaliningrad	Yes	No	Yes	No	No
Arkhangelsk	Yes	Yes	No	No	No
Ivanovo	Yes	Yes	No	No	No
Rostov	Yes	No	Yes[a]	No	No
Tyumen	Yes	Yes	No	No	No

NOTE: a) Under development.

SOURCE: I. Bashmakov and A. Cleveland, *Energy Efficiency in the Regions of Russia* (Moscow: CENEf, 1997; revised, Spring, 2000). See also www.glasnet.ru/~cenef/regions.htm.

in 1994, and in new buildings saved 20 percent compared to similar ones built just prior to passage of the Moscow code. Moscow and other cities will cut unit consumption by another one-fifth. Further reductions will be made when a third-stage standard is imposed beginning in the year 2000.[45]

Russia's local governments have paid general attention to energy saving and have created programs to capture some of the potential. The motivation is pecuniary: Residents typically paid one-quarter or even less of the cost of housing and utilities, and heating accounted for most of that cost. The Russian reform leader Boris Nemtsov in 1997 outlined a plan to impose full cost-recovery on households by the year 2003, but the plan met heated opposition in parliament.[46] Only six or seven of the numerous regional governments have actually allocated budgetary

Figure 2.4   Infrared View of Hospital in Troitsk, Russia, 1998.
NOTE: Surface temperature in degrees Centigrade. Bright and dark
correspond to high and low rates of heat loss.
SOURCE: Center for Energy Efficiency, (CENEf), Moscow, 1999.

funds for efficiency activities or imposed local codes and standards.
Under Russian law, building codes and standards are provided at the
national level only as a model for acceptance or rejection by the local
governments.[47]   Russian multifamily buildings have long been constructed of prefabricated concrete panels. These have poor thermal
properties, and implementation of the codes and standards, and the
need to save heat costs, has prompted a change in the panel construction industry around Russia. An infrared picture of a hospital, for example, shows an uneven distribution of heat in the building and significant heat losses from the upper stories (see Figure 2.4). At this writing,
manufacturing plants capable of producing three-layer panels with improved thermal performance now operate in over fifty-seven of Russia's ninety-eight regions.[48]

A number of prior efforts have been made to reduce the problem of
energy waste in buildings.[49] A demonstration project in Krakow in the
early 1990s showed that valves, controls, and heat exchangers, along
with simple measures such as weatherstripping around the leaky windows, could save over one-quarter of the energy consumed by residential buildings. This level of improvement was achieved with less than a
four-year payback, and, significantly, was achieved without changing
the economic behavior of the consumers. Savings on the order of 20 percent have been achieved in demonstration projects for apartment energy
efficiency in Russia.[50] In other words, the demonstration showed the potential for a technical fix, not a behavioral or economic response

prompted by raising prices. Price signals are, of course, essential to make efficiency work in the residential sector, and metering equipment and price reform are essential to capture the technical potential indicated by demonstration projects.

In response to the need for residential energy efficiency, the World Bank developed a $450 million program in six cities under the "Enterprise Housing Divestiture Program" (EHDP), the largest energy efficiency lending package in its history.[51] Bank economist Dennis Whittle was seeking a solution to a major sticking point in economic reform.[52] Enterprise housing had to be divested to free industry from the burden of maintaining the buildings, including paying for utilities, and to free residents from the obligation to a firm that came with living in the firm's flat. A solution was to turn ownership over to the residents of the flats. Residents would then be able to keep, trade, rent, or sell housing, and provide a mobile and independent labor force, at the same time ridding the enterprises of the crushing overhead. The sticking point, however, was that the new owners would not be able to afford the utilities. Experience in Poland and estimates in Russia indicated that the heat and power to run the flat would cost one-fifth or more of total household income. On average, a family in Russia was making only about $150 per month during the 1990s.

The World Bank efficiency project faced technical, behavioral, and institutional difficulties. The apartments had no heat meters, and heat rates charged on the basis of the size of flats were unrealistically low. Utilities would have to triple consumer bills to recover actual costs. If residents' flats were metered and prices were raised, they would be expected to save money by turning down or off radiator valves while they were out or sleeping. They could plug air leaks around windows or through cracks in the walls. But of course the residents lacked not only meters to count their energy use—and thus reward them for saving—but valves as well. At the time, no one in Russia manufactured quality caulk or weatherstripping.

Two-thirds of the financing was provided by a loan from the World Bank itself, and one-third by contributions from the participating Russian cities. The program would have been difficult simply due to the problem of procuring numerous pieces of equipment for 500,000 flats in six different cities. It was made much more so by the need to meter heat and to collect payments from the residents to pay off the World Bank loans. A metering system had to be created and a much stricter payment system imposed. Previously, residents paid their heating bills on the basis of the number of square meters of flat they occupied—that is, based on the size of their apartments. Now they would be responsible for their heating bills and would have an incentive to close windows and shut off valves when the heat was not needed. Metering of individual flats, however, is very difficult to do accurately without incurring exorbitant fees. This problem, however, is a common one in Western Europe, and satisfactory allocation solutions using low-cost thermal monitors, sometimes called "cost allocators," have been devised. Estimated annual energy savings in the buildings will total 30 percent as a result of the program, with all measures to pay for themselves in five years or less.

In the course of developing the program, it became obvious that district heating upgrades may not always be the most efficient investment. Natural gas is available in much of Russia, of course, and the question arose as to whether heat and hot water could be provided with decentralized boilers located at each block of flats. A demonstration project of this type has been developed using Global Environment Facility funding in Ryazan in conjunction with the EHDP program.[53]

The "Saudi Arabia of Efficiency" seemed to get serious about efficiency under the leadership of former first deputy prime minister Boris Nemtsov. Working with former foreign minister Andrei Kozyrev, new programs were implemented to finance municipal energy efficiency.[54] Upheaval in the Russian central government may matter less for continuation of this policy than might be expected. The motivation for leadership for efficiency naturally comes at the local level, where leaders must confront their constituents when there is no heat, when there is no money for the city's largest employer to pay its gas bill, or when power is turned off because consumers have not paid and fuel suppliers will no longer advance supplies. This problem is most acute in remote areas like Sakhalin, where heat and power often are unavailable in the coldest months.[55] Leadership to solve these problems must come from within a country, and the most likely place to find such leadership, now and in the future, is at the local level.[56]

Ukraine

Energy demand reduction is both an economic and a security concern—a painful one—for Ukraine. Not only is energy use an essential element of industry restructuring, but energy supply is threatened because the country imports half the primary energy it consumes. Moreover, it obtains the great majority of its imports from a single supplier, Russia.

Ukraine has endured a reputation as a place unfriendly to foreign investment. The World Bank and various donor nations, including the United States, have at various times threatened to cut off assistance unless progress was made in reform. As in Russia, the World Bank has cited instances of corruption or favoritism in the privatization of firms. The western press has highlighted problems with contracts in the lucrative natural gas transmission and distribution sectors having gone to close colleagues of high government officials. Corporatization of the energy sector has been stymied, although efforts have been made to create a wholesale spot market for electric power production. On the other hand, Ukraine has raised prices to world levels, and did so far faster than the United States reformed its own fuel prices during the energy crises of the 1970s and early 1980s. Nonpayment of utility bills remains a problem, but the problem is not peculiar to the energy sector.

Ukraine was the first country of the region to establish a ministry-level committee to promote energy efficiency. Through the leadership of energy specialist Mykola Raptsun and gas expert Mykhailo Kovalko, this committee instituted a government-wide plan to implement efficiency and developed financial resources to help Ukraine adjust to the high cost of imported energy.[57]

An early effort to promote efficiency through the use of funds collected by means of a special energy tax was defeated by the energy industry. A 1996 government decree would have created the fund with a 15 percent tax on retail gasoline and diesel fuel. The goal was to raise $50 million for energy conservation. The decree was revoked, however, after energy companies staged a strike to protest the measure.[58]

Some progress has been made in demonstrating how to finance energy efficiency in this difficult context. A World Bank project to retrofit institutional buildings in Kyiv has been designed, and a demonstration project in a small set of those buildings suggested savings of 26 percent could be expected, just by installing controls to turn down the heat when the buildings were unoccupied overnight and on weekends. A private sector prototype project at the Gostomel Glass company, which makes bottles for Coca-Cola and the popular Obolon brewery, used western assistance to help persuade private investors from Europe to finance a new glass furnace that will consume only 40 percent as much natural gas per bottle manufactured.[59] The European Bank for Reconstruction and Development had developed plans to finance an energy service company, and Urkgasprom, the Ukrainian natural gas company, has expressed strong interest in replacing inefficient compressor turbines on its pipelines. The turbines pump gas at an efficiency of 25 percent, compared to new U.S. and European systems that exceed 40 percent efficiency. The progress made in developing these investment opportunities suggests that Ukraine is capable of getting its demand-side house in order, and that in the coming years, the nation will enjoy a substantial business in this effort.

Other Countries

Across the region, price reform has varied, especially in the energy sector. Hungary, despite being an "A" student, has raised industrial natural gas prices less than every other major energy-consuming country of the region except Russia, and its residential gas prices are lower still.[60] Industrial electricity in 1998 cost less in Hungary than in Ukraine. Similarly, privatization has not been consistently applied, even in the better-off nations. Privatization and inflation appear to be closely, and inversely, linked, meaning that privatization results in less inflation.[61] It would be difficult to establish a statistical correlation between privatization and economic recovery, however. Poland, after all, has moved slowly to privatize, above all in the energy sector, and Russia would be a success story, if ending state control of firms were all there is to transition.

Frontier Financing

Financing in the 1990s was a big hurdle for transition economy energy efficiency. A vivid illustration is the Russian municipal heat crisis. Residents unwilling or unable to pay for the heat they used or to make investments to save energy had by 1998 driven cities across Russia nearly to bankruptcy and raised the specter of social disaster.[62] One-third of municipal budgets went to pay for heat subsidies. At best, the crisis represented a sticking point in price reform and privatization efforts. At worst, millions of Russians would soon find themselves without heat and hot water in the frozen north.

Chelyabinsk typifies the cities' plight. Like most Russian cities, it is dominated by a huge enterprise, a steel mill. The mill's power plant made electricity for the mill and heat for apartments in town. Privatization required the mill to survive by selling its steel, which it exported. To cut costs, the plant laid off workers and divested itself of the workers' housing and heating system. The city began collecting heat payments from the residents and paid the factory for the heat. A utility commission raised prices for the heat to its economic value. The city, in the meantime, was obligated to pay the difference between residents' heat bills and the cost of the heat. Consequently, 35 percent of Chelyabinsk's budget went to subsidize heat.

One way out of this trap was to privatize the heating utility. However, selling a bankrupt, decrepit heating system with nonpaying customers seemed optimistic. Another option was to shut off heat to those who did not pay, but residents lived in multifamily housing without individual heat meters and without individual controls on their flats. Technically and politically—Chelyabinsk is at the edge of cold Siberia—this approach seemed to be unfeasible.

A third option was to invest in conserving heat. Three basic problems constrained that investment, and they are typical for energy savings investments across the region: lack of equity, lack of customer finance, and lack of access to capital markets. In a normal economy, a financier may think of a financial problem simply enough as one of "the three Cs."[63] Finance is, at bottom, based on extremely simple principles, including those of character, cash flow, and collateral. The question of character is simple enough to resolve in the U.S. context. A credit agency is engaged for a fee and determines whether the loan applicant has "credit." That characterization requires knowing whether the borrower has borrowed money in the past and repaid loans on time and has ability now to pay. But if the loan applicant lives in an economy in which lending for private profit was a crime against society, he or she is unlikely to have had the opportunity to have earned "credit."

The second C is easier, if more expensive, to address. Cash flow can be estimated for energy efficiency projects in terms of reduced utility bills. This approach is not uncommon in the west, where it is necessary to do a business plan, feasibility study, or "appraisal" to demonstrate with due diligence that the cash flow estimates are meaningful and that the lender should have confidence in them. Preparation of such plans is both a cultural and a technical problem in the former communist countries. It is a cultural problem because few persons have experience in preparing them. Educational efforts are needed before many business plans can be developed. It is a difficult technical problem because there is almost never any baseline of energy use against which to make the savings projections. District heating systems often lack even the most basic metering system, which is to say that no one even knows how much energy leaves the heating plant, and certainly not how much heat is consumed in an individual apartment. Modeling by experienced engineers can be used to generate savings estimates, however, and has been applied in major World Bank projects.

The third C—collateral—represents the problem of security. If an American wants to buy a house, all that is required is a credit check (does he or she have character?), salary verification (is cash flow sufficient?), and if all else fails, how much the house is actually worth should the mortgage be foreclosed. Collateral would work fine for securing energy efficiency projects in the region, except that no one knows for sure what commodities or property are worth and whether the legal and political infrastructure would permit the lender to recover unpaid loans by seizing assets.

Efficiency's difficulties reflect the financial immaturity of transition economy markets. Industries, local governments, and apartment dwellers in the transition economies lacked—and still lack—the ability to pay up front. Vendors usually are unwilling to finance customers' purchases, and lenders usually are unwilling to finance transactions without a guarantee of repayment. These problems are compounded by the "small project" problem, which is to say that because productivity projects such as efficiency investments tend to be much smaller than the mega-deals in the oil and power industries, transaction costs can overwhelm the deals. Management costs for loans are relatively fixed, as are due diligence requirements. Therefore, the cost of developing a $5 million municipal district heating or a $15 million industrial cogeneration project may be financially attractive, but the up-front costs are likely to be the same as for a $100 million power plant or $1 billion gas or oil deal. The risk-to-reward ratio tends to be high enough to send financiers looking elsewhere for investments.

As the regions' economies mature, as financial reforms progress, and as markets become more creative in solving these problems, efficiency investments will slowly but surely be made. Local—and lower-cost—consultants will deal with issues of character, cash flow, and collateral. Local banks will increasingly find it in their interest to help their customers, including municipalities, remain solvent, and loans to them or provision of guarantees for loans to them will become a good business. These banks will have the knowledge necessary to determine whether the loan is likely to get repaid.

A solution once sought to the collateral problem is the sovereign guarantee. The World Bank, as a requirement of its charter, must have a sovereign guarantee—a guarantee by the treasury of the central government—that the loan will be repaid. Obtaining money on financial markets is cheap for the World Bank because no one has ever defaulted on a World Bank loan. However, the World Bank can finance only a tiny fraction of the region's needs. The rest must be financed by the private sector. Although private players once sought sovereign guarantees for deals in the region, they have become much more difficult to obtain and, in the case of Russia, worthless. Moreover, the sovereign guarantee involves having government pick the winners, picking the projects to get financed, an approach diametrically opposed to the reform needed in the region.

One creative solution receiving considerable attention is the energy service company (ESCO) approach. ESCOs are private firms that provide financing, technical consulting, installation, and management services to customers needing efficiency improvements. ESCOs represent a significant and growing business in France, where they were invented, and in the United States, where they commonly serve industry, commercial buildings, and public institutions. In past practice, an ESCO would make an investment and share the energy savings with the customer. Reduced energy payments would generate cash flow for paying off the investment, providing a profit to the ESCO while saving the customer money. After a few years, all the savings belong to the customer. This "shared savings" approach, however, has been replaced with performance contracting, in which the ESCO guarantees savings of a given amount. This approach gives the customer greater certainty for budget planning and gives the ESCO greater incentive to perform. If the expected savings are not achieved, the ESCO must pay the difference. Allowances are made in the contracts, of course, for variables such as extreme weather and increased operation or production schedules.

Some effort has been made to introduce ESCOs into Eastern Europe and the former Soviet Union. A success story comes from the Czech Republic, where SEVEn first became interested in the ESCO approach and invited a foreign advisor to explore the possibility of introducing ESCOs into what was then Czechoslovakia (see Table 2.4).[64] Subsequently, SEVEn introduced the U.S. firm EPS to several potential customers. Six months after EPS's first visit to Prague, the company had obtained several million dollars in contracts and implemented them profitably, to the apparent satisfaction of its customers, which included industry, hospitals, and municipal institutions. A key to success in this case was the provision of a loan guarantee to the ESCO by a major equipment manufacturer, which wanted to capture market share in the region.[65] Backing was later provided by a U.S. utility seeking international business. This alliance created problems for the firm, however, when the 1998 financial crisis frightened the utility out of the international market.

The International Finance Corporation (IFC) and the European Bank for Reconstruction and Development (EBRD) have attempted to deliver energy efficiency in the region by creating local energy service companies. These two banks have provided equity investment. In this way, ESCOs have been created in the Czech Republic, Hungary, Poland, and Ukraine. Significantly, the terms provided by the IFC and EBRD for private sector participants in these ventures have been almost identical to commercial banking requirements. Typically, the banks would put up a minority stake, with a private western firm taking the biggest share, but also involving a local investor. The development banks thus have tried not to distort the market significantly. This approach, however, has slowed their ability to move quickly and aggressively to support efficiency and therefore to accomplish their reform objectives. Negotiations to create ESCOs in Russia and Ukraine were especially slow and difficult.

A different approach to financing has involved the creation of lines of credit through private banking institutions. The IFC, for example, in the early 1990s made several million dollars of funds available for productivity improvement to Polish industry through the Polski Bank Rozwoju (PBR), the Polish Development Bank. This private bank was well-managed and profitable, but its interest in promoting and processing IFC loans was reduced by the obligation to guarantee repayment and to cover all transaction costs. PBR agreed to charge a fee of about 2 percent,

TABLE 2.4   ESCO Projects Completed in the Czech Republic

Facility	Energy Savings	ESCO
Elementary School # 1, Litvínov	32 %	SUE, Ltd.
Elementary School, Beèov u Mostu	62 %	SUE, Ltd.
T. G. Masaryk City Hospital, Jilemnice	76 %	EPS ÈR, Ltd.
Bulovka University Hospital, Prague	42 %	EPS ÈR, Ltd.
Elementary School, Obrnice u Mostu	29 %	SUE, Ltd.
High School, Chomutov	42 %	SUE, Ltd.
Elementary School # 3, Chomutov	44 %	SUE, Ltd.
Elementary School, Litvínov-Janov	34 %	SUE, Ltd.
Krušnohorská Polyclinic, Litvínov	52 %	SUE, Ltd.

SOURCE: "EPC Projects in the Czech Republic 1992-97," *News at SEVEn*, 6, no. 1, (April 1998): 1.

making investments in the Polish stock market, financing foreign trade, or even buying sovereign debt more profitable. Borrowers, moreover, were required to take the foreign currency risk of the loan because the funds had to be repaid in dollars, and although this risk could be managed through currency hedging, it increased loan costs. More important, would-be lenders were discouraged by onerous and time-consuming loan application requirements and an exceedingly slow approval process. Potential borrowers cited these delays as their main reason for not utilizing the funds. The PBR fund, like similar ones across the region, went essentially unsubscribed. Attempts to create specialized energy efficiency funds were also made elsewhere, including the Czech Republic, but generally have not met with success.[66] Lack of private equity constrained the European Bank for Reconstruction and Development from contributing to such funds.

The role of the European Bank for Reconstruction and Development has been bogged down in part by its other obligations in the region, especially regarding its Nuclear Safety Account. The Bank had succeeded in placing only $4.1 billion of investment in the region by mid–1997.[67] One issue is its practice of sharing project risk with commercial financiers. Typically, EBRD finances only one-third of a total project amount, the remainder coming from international commercial banks or investors. This approach avoids crowding out commercial finance, but because meeting commercial lending standards is extremely difficult in the region, EBRD has limited flexibility.

Perhaps the most contentious aspect of energy efficiency finance is the disproportionately larger amount of funding that flows to energy supply projects. Such loans, typically provided by the trade promotion organizations of the United States and Europe, can in the absence of balance with the demand side be seen as subsidies for consumption. In its worst characterization, the trade agencies are said to have filled the void left by the defunct planning agencies. What is clear is that the flow of funds is unbalanced. Very little money flows to demand-side investments. Yet, the U.S. Ex-Im Bank facilitated creation of a billion-dollar escrow account for the sale of U.S. oil and gas equipment to Russian firms.[68] Ex-Im provided the Czech utility CEZ a loan of $350 million to complete the Temelin nuclear power plant. When environmentalists have criticized Ex-Im of bias favoring supply-side projects, the agency has responded that is not a development agency but is "passive" in that it provides financing to companies that come to it seeking financing. Ex-Im also points out that it placed a line of credit for $15 million at the Bank Ochrony Srodowiska (Bank for Environmental Protection) in Poland for environmentally friendly investments.[69]

Supporters of the energy efficiency industry often lament its decentralized nature. Because the technology and services of efficiency affect every aspect of the economy, there is a large disparity between the importance of efficiency and its impact on policy. This disparity is natural given the nature of politics. Policy makers are inclined toward "shiny objects," projects that one can visit, point to, get voters excited about. It is hard to have a ribbon-cutting ceremony for a steam-trap project. But a more serious political problem is that the manufacturers and service providers for energy efficiency are numerous, competitive, and of divergent interests. Building effective lobbies can be difficult in such an industry.[70]

Consumer interests represent a reasonably coherent interest group, but if consumer groups support price reduction rather than price reform, the interest of energy efficiency is not served. Two organizations, however, have proven the value of consumer organizations. Perhaps the first industrial energy consumer organization, the Association of Energy Managers, was created in the former Czechoslovakia by Czech entrepreneur Ivo Slavotinek. The first such Russian organization, the Russian Energy Managers Association (REMA), founded by Mikhail Berner, formerly energy manager for the Moskvitch automobile manufacturing company, has supported energy reform, particularly in utility regulation. By insisting on transparency in utility rate making and by supporting financial policies that support efficiency, the group has provided leadership in helping Russia make the transition to a market economy.[71] High energy costs remain a problem for Russia's recovery and for energy reform, because only about one-seventh of Russian consumers pay their bills.[72] The nonpayments problem is far less severe in other nations, including the Czech Republic, Hungary, and Poland, which suffered comparable price increases.

Conclusions

Governments, both within the region and outside, have missed opportunities to save energy and, in so doing, facilitate reform, promote economic growth, and reduce environmental pollution. The most egregious error is in the continuation of energy subsidies. Local governments mandate cross-subsidies and direct incentives for supply development. They fail to impose hard budget constraints. Foreign governments, unfortunately, baldly promote supply-side constituents, and in so doing, subsidize their own companies seeking business in the region. Bilateral lending institutions in particular have proven unbalanced in their approach.

Although Central Europe has successfully reduced its energy wastefulness, primarily as a result of macro-economic reform, targeted energy efficiency programs have become increasingly important for cost efficiency and environmental protection. These will grow more important as the region's economies grow. Successful efficiency programs have shared five key elements, including:

- Focus on making markets work
- Reliance on local expertise
- Matching funding or self-financing requirements
- Emphasis on high internal rates of return
- Attention to financial guarantees

Government assistance has been most helpful when it has leveraged private business interests or at least leveraged financial investments. For example, some programs have created investments totaling $50 for every dollar expended.

The cost of energy waste remains an overwhelming problem for the region's consumers. The crises in the region's cities and factories will not soon go away. The lost wealth and the unnecessary pollution due to energy waste may long continue. Some western efficiency programs, however, do provide enlightened assistance to the region with programs that help ordinary people adjust to the realities and burdens of energy reform. Energy efficiency programs clearly help consumers cut costs and pollution and improve their economic prospects. Conservation programs carefully targeted for local governments, real consumers, and small business make the most difference.

Notes

1. Russian homes averaged about 3.5 persons and 63 square meters per household in 1995. Eric Martinot, *Investments to Improve the Energy Efficiency of Existing Residential Buildings in Countries of the Former Soviet Union*, Studies of Economies in Transformation, no. 24 (Washington, D.C.: World Bank, 1997), 8.

2. Just as one large cornflakes box uses less cardboard than two small ones holding the same amount of cereal, a large apartment building has less external building envelope area than several single-family houses providing the same amount of floor space. That is, multifamily housing has a lower surface-to-volume ratio than single-family housing and therefore smaller energy losses per unit of living space.

3. Igor Bashmakov, Director, Center for Energy Efficiency (CENEf), personal communication, Moscow, 1991.

4. The value is 220–260 kJ per square meter per degree-day. Sławomir Pasierb, President, Polish Foundation for Energy Conservation, personal communication, Katowice, Poland, October 1997.

5. Eric Martinot, *Investments to Improve the Energy Efficiency of Existing Residential Buildings in Countries of the Former Soviet Union*, Studies of Economies in Transformation (Washington, D.C.: World Bank, 1997), 10.

6. Richard C. Paddock, "Moscow Heating Pipes Create Lethal Traps Safety: Sinkholes Caused by Hot-water Leaks Result in Grisly Deaths. Utility Lacks Repair Funds," *Los Angeles Times*, 7 April 1998, p. A1.

7. United Nations Economic Commission for Europe, *An Efficient Energy Future: Prospects for Europe and North America* (London: Butterworths, 1983), 33–37.

8. John H. Gibbons and William U. Chandler, *Energy: The Conservation Revolution* (New York: Plenum Press, 1981), 161–165.

9. S. Meyers, L. Schipper, and B. Lebot, *Domestic Refrigeration Appliances in Poland: Potential for Improving Energy Efficiency*, Lawrence Berkeley Laboratory (Berkeley, Calif.: August 1993), 5–7 (citing unpublished data compiled by the Polish Foundation for Energy Efficiency, Warsaw).

10. U.S. Congress, Office of Technology Assessment, *Energy Efficiency Technologies for Central and Eastern Europe* (Washington, D.C.: Government Printing Office, 1993), 16.

11. Lee Schipper and R. Caron Cooper, *Energy Use and Conservation in the USSR: Patterns, Prospects, and Problems* (Berkeley, Calif.: Lawrence Berkeley National Laboratory, 1991), 5.

12. International Energy Agency, *Energy Policies of Poland, 1994 Survey* (Paris: Organisation for Economic Cooperation and Development, 1995), 57–58.

13. William Chandler, *The Changing Role of the Market in National Economies* (Washington, D.C.: Worldwatch Institute, September 1986), 20, citing data provided by the International Iron and Steel Institute, Brussels, and Maurice Y. Meunier and Oscar de Bruyn Kops, *Energy Efficiency in the Steel Industry with Emphasis on Developing Countries* (Washington, D.C.: World Bank, 1984).

14. OTA, *Energy Efficiency Technologies*, 46.

15. Ibid.

16. International Energy Agency, *Energy Policies of Ukraine: 1996 Survey* (Paris: Organisation for Economic Cooperation and Development, 1996), 129.

17. OTA, *Energy Efficiency Technologies*, 46.

18. See Organisation for Economic Cooperation and Development, *Transportation Statistics 1997 Annual Report*, (Paris: Organisation for Economic Cooperation and Development, 1997), 236.

19. Whether they could move, however, is a different matter.

20. OTA, *Energy Efficiency Technologies*, 46.

21. SEVEn, *Case Studies of Technology Transfer in the Czech Republic* draft (Prague, January 1997).

22. OECD, *Transportation Statistics Annual Report 1997*, 236–258.

23. Sławomir Pasierb, President, Polish Foundation for Energy Efficiency, personal communication, Katowice, Poland, 2 October 1998.

24. Goscomstat, *Налияие Легковых Автомоьилей вСобствинности Граждан по Гегиогам Российской Федерации* (Moscow: Russian Federation, 1997), 101.

25. OECD, *Transportation Statistics Annual Report 1997*, 244–245.

26. Ibid., 251–252 and 257–258.

27. Janusz Michalik, Sławomir Pasierb, Jerzy Piszczek, Michał Pyka, and Jan Surówka, *Evaluation of the Feasibility and Profitability of Implementing New Energy Conservation Technologies in Poland*, Polish Foundation for Energy Efficiency (FEWE) (Katowice, Poland, 4 May 1993).

28. IRR stands for internal rate of return, the rate at which the discount rate would have to be set for the investment to provide a zero rate of return. It is a standard measure of return on investment.

29. See *Cogeneration and Energy Efficiency at Avdeevka, 1997*; *Energy Efficiency at Gostomel, 1997*; and *Financial Review and Appraisal of the Economic Viability of the Open Joint-Stock Company Avdeevka Coke Chemical Plant, 1998*, all available from the Pacific Northwest National Laboratory at www.pnl.gov/aisu.

30. Meredydd Evans, J. W. Currie, N. V. Raptsoun, J. D. Duane, and K. E. Nelson, *Energy Efficiency Business Opportunities in Ukraine*, PNNL–10374, Pacific Northwest Laboratory (Richland, Wash., March 1995).

31. Henryk Gaj and Maciej Sadowoski, *Climate Change Mitigation Case Studies from Poland*, edited by Susan Legro, Pacific Northwest National Laboratory (October 1997).

32. Ustina Markus, "Energy Situation in Ukraine," *OMRI Daily Digest* II, no. 9 (14 January 1997); http://www.rferl.org/newsline/1997/01/3-CEE/cee-140197.html.

33. Energy Law of Poland, published 10 April 1997, translated from Polish by the Polish Foundation for Energy Efficiency, Warsaw.

34. Michalik et al., *Evaluation of the Feasibility and Profitability of Implementing New Energy Conservation Technologies in Poland*.

35. Gaj, Sadowski, and Legro, *Climate Change Mitigation Case Studies from Poland*.

36. International Energy Agency, *Energy Policies of Poland* (Paris: Organisation for Economic Cooperation and Development, 1995), 161–162.

37. S. Pasierb, K. Niedziela, and J. Wojtulewicz, "Structural and Technological Changes of GHG Emissions During the Transition Period," *Polish Industry Environmental Management* 20 (1996): 1; and S. Pasierb, *Poland's Alternative Energy Policy Until the Year 2020* (Katowice, Poland: Polish Foundation for Energy Policy, 1996), 5–10.

38. Jiri Zeman, "Energy Policy in Transition: A Tool That Promotes and Discourages Energy Efficiency," in *Examples in Action: Sustainable Energy Experiences in Developing and Transition Countries* (Washington, D.C.: International Institute for Energy Conservation, 1996), 1–6. Reprint available from SEVEn, Prague, Czech Republic.

39. Ibid. See also Office for Standardization and Metering, *Czech Technical Norm On Standards for Thermal Insulation of Buildings*, CSN 73 0540:94 (Prague, Czech Republic, 1994).

40. United Nations Economic Commission for Europe, *An Efficient Energy Future: Prospects for Europe and North America* (London: Butterworths, 1983), 34.

41. Jaroslav Marousek, Martin Dasek, Bedrich Schwartzkopf, and Marie Havlickova, *Climate Change Mitigation Case Studies from Russia, Poland, Czech Republic*, edited by Susan Legro, Pacific Northwest National Laboratory (October 1997).

42. Jaroslav Marousek, SEVEn, the Energy Efficiency Center, personal communication, Prague, October 1998.

43. International Energy Agency, *Energy Policies of the Russian Federation* (Paris: Organisation for Economic Cooperation and Development, 1995), 28–31.

44. See "Energy Efficiency in Russia's Regions," Center for Energy Efficiency, Moscow, April 2000, http://www.glasnet.ru/~cenef/regions.htm.

45. Yuri Matrosov, David Goldstein, and Mark Chao, "Implementation of Regional Building Energy Codes in Russia: Prospects for Imports and Indigenous Production of Advanced Technologies," *Energy Efficiency* 18 (January-March 1998): 2–6.

46. "Russians to Pay More for Heating," *RFE/RL Daily Digest*, 29 April 1997, http://www.rferl.org/newsline/1997/04/1-RUS/rus-290497.html; Michael R. Gordon, "Yeltsin Attacks Soviet-Era Housing Benefits," *Washington Post*, 13 July 1997, p. 1.

47. The Russian federal model code for buildings thermal performance is referred to as SNiP II–3–79 (1995) "Thermal Engineering." The recommended standards for the year 2000 are said to be comparable to standards in Canada and Sweden. See Matrosov et al., "Implementation of Regional Building Energy Codes in Russia," 2–6.

48. Matrosov et al., "Implementation of Regional Building Energy Codes in Russia," 2–6.

49. Tom Secrest, Andrew Popelka, and Jirki Zeman, *Assessment of the Buildings Sector Efficiency Resource for the City of Plzen, Czech Republic*, PNNL–10898, Pacific Northwest Laboratory for the U.S. Department of Energy (1996); Tom Secrest and Andrew Popelka, *Integrated Assessment of Supply and Efficiency Resources for the District Heating System, City of Handlova, Republic of Slovakia*, PNNL–11205, Pacific Northwest Laboratory for the U.S. Department of Energy (1996); Tom Secrest et al., *Polish-American Children's Hospital in Krakow, Poland*, PNNL–8544, Pacific

Northwest Laboratory for the U.S. Department of Energy (1993); Jirka Zeman, Tom Secrest, and Andrew Popelka, *Cesky Krumlov Energy Resource Assessment*, Pacific Northwest Laboratory for the U.S. Department of Energy (1993).

50. Eric Martinot, "Investments to Improve the Energy Efficiency of Existing Residential Buildings in Countries of the Former Soviet Union," in *Studies of Economies in Transformation* (Washington, D.C.: World Bank, 1997), 26.

51. A very brief description of this project can be found online at the Moscow Center for Energy Efficiency, CENEf, site: http:/www/glasnet.ru/~cenef. See G. Glukhova (Foundation for Enterprise Restructuring, Government of the Russian Federation), "Financing Scheme for the World Bank's Enterprise Housing Divestiture Project," *Energy Efficiency* (April-June 1996): 10–11.

52. The reader should be aware that the author was involved in the design and development of this project.

53. Martinot, "Investments to Improve the Energy Efficiency of Existing Residential Buildings," 35. See also Battelle, *Consultant's Final Report on Project Preparation Activities for the Enterprise Housing Divestiture Project*, Pacific Northwest National Laboratory, consultant's report for Ministry of Economy of the Russian Federation (Moscow, 1996).

54. Andrei Kozyrev in 1998 served in the Russian Duma, the lower house of parliament, representing Murmansk. He served on a committee concerned with the problems of the Russian north, and remains very active on consumer energy issues, including fuel shortages and district heating efficiency. Andrei Kozyrev, personal communication, Moscow, 13 October 1998..

55. "Lights Out in Sakhalin," *RFE/RL Daily Digest*, 9 April 1999, http://www.rferl.org/newsline/1999/04/090499.html.

56. See William Chandler, Alexei Makarov, and Zhou Dadi, "Energy for the Soviet Union, Eastern Europe, and China," *Scientific American* (October 1990): 127. See also *First National Communication of the Russian Federation Under the United Nations Framework Convention on Climate Change* (Moscow: Interagency Commission on Climate Change, 1995); *Energy Strategy of Russia: Major Provisions* (Moscow: Interagency Commission for the Coordination of Programme Development, 1995) (as reproduced in *Energy Policies of the Russian Federation*); I. Gritsevich, "Concept of Russian Federation Transition to Sustainable Development," *Energy Efficiency* no.10 (1996); CENEf, *Russian Energy Picture* (Moscow: CENEf, July-September 1996); I. Bashmakov, V. Zhuze, Yu. Dashevsky, and S. Sorokina, *Improving District Heating Efficiency in the Metallurguichesky District of Cheliabinsk*, CENEf (Moscow: CENEf, 1996); B. Baranov, "Automation Control System and Electric Power Quality Check at Moscabelmet," *Energy Efficiency*, 1, no. 3 (1994): 3; I. Bashmakov, I. Gritsevich, and S. Sorokina, *System of Institutional Measures to Stimulate Energy Efficiency in Russia as a Part of the GHG Mitigation Strategy*, CENEf (Moscow, 1996).

57. Translation of State Committee on Energy Savings (Mykhailo Kovalko, Chairman) *Energy Efficiency Five-Year Plan*, as approved by the Council of Ministers (Arena-Eco, Kyiv, 1997).

58. Sergei Avdiushin et al., *Cities in a No-Regrets Climate Strategy: Lessons from Transition Economies*, PNNL-15046, Pacific Northwest National Laboratory (October 1997) (citing contributions by the Ukrainian authors).

59. Steven A. Parker, Valery Maschenko, Meredydd Evans, and Vladimir Laskarevsky, *Energy Efficiency at Gostomel: Recommendations and Energy Audit Report for Gostomel Glass Plant*, Pacific Northwest National Laboratory (Washington, D.C., October 1997).

60. See International Energy Agency, "Statistics," at www.iea.org/stats.

61. This conclusion is based on simple regression analysis using IMF and World Bank data.

62. Consumers can neither conserve nor become more efficient in the current situation. They cannot simply turn off the heat—they have no valve. They could install heat meters, valves, controls, etc., but they have no money, and it takes a steady hand to sign loan papers to people in such conditions.

63. Thanks to David Smith, independent financial consultant, for suggesting this teaching device in a seminar at Battelle, Pacific Northwest National Laboratory, July 1995.

64. Dan Guttman, "Energy Performance Contracting: Legal and Commonsensical Considerations" (paper presented at Energy Efficiency Business Week, Prague, 11 November 1993).

65. "Session on Energy Performance Contracting," in *Energy Efficiency Business Week 1995* (Prague: Czech Center for Energy Efficiency, SEVEn, 1995). See also M. Diduåková and M. Votápek, *Jak Investovat do Úspor a Šetřit Vlastní Kapitál* (Prague: SEVEn, 1995).

66. "*Financial Sources for Energy Efficiency*," Study #95/002/c, SEVEn, the Czech Energy Efficiency Center (Prague, 1995).

67. European Bank for Reconstruction and Development, Internet home page, March 1998, http://www.ebrd.com/opera/index.html.

68. See "Russian Oil and Gas Framework Agreement," U.S. Export-Import Bank, http://www.exim.gov/country/ebd-y-18.html.

69. Ex-Im's focus on trade, as opposed to policy, was made clear in a political dispute involving the bank's guarantee of a $187 million loan for purchase of combines by a Ukrainian state-controlled agricultural collective, ostensibly at the cost of private Ukrainian farmers. See Jane Perlez, "Capitalist Tools Benefit Ukraine Collectives," *New York Times*, 15 July 1996, p. 1.

70. Industrial restructuring may have reduced competition in industrial and building controls when Electrowatt of Switzerland bought the electronic controls manufacturer Landis & Gyr. (Concerning the results of this merger, see William Hall, "Electrowatt Announces Restructuring," *Financial Times*, 1 October 1996.) Landis, along with Honeywell and Johnson Controls, was among the more influential energy efficiency companies aggressively seeking business in Central Europe and the former Soviet Union.

71. The Russian Energy Managers Association was founded by Berner and Igor Bashmakov, director of the Russian Center for Energy Efficiency (CENEf). Bashmakov requested, and obtained, funding from the regional AID office in Moscow. The energy program at AID in Moscow at the time was headed by Gene George.

72. Chrystia Freeland, "Moscow Mayor Leads Call for Energy Price Cuts," *Financial Times*, 4 November 1996, p. 18.

3

Nuclear Safety and Security

Chernobyl[1] and the threat of repeating the 1986 catastrophe at that nuclear reactor dominate the international agenda for regional technical cooperation. Half of all western technical assistance to the region has been allocated to nuclear power safety programs. The Group of Seven industrial nations have identified three dozen Soviet-designed reactors as inherently dangerous due to design flaws. Although those plants provide less than 2 percent of the region's energy and could, with a few exceptions, be closed without hardship, regional leaders have declined to close them. Western nuclear specialists, unable to influence this decision, have acquiesced and work instead to ameliorate immediate hazards such as fire and operator error. Alvin Weinberg, the U.S. mathematical bio-physicist who helped develop power reactors, once described nuclear power as a Faustian bargain. Western governments, still hoping to avert a second Chernobyl, have concluded that they cannot compel transition economies to close down their unsafe reactors, so they must instead introduce a "safety culture" to reduce the risk of a new catastrophe. Because older models of Soviet-designed reactors have fundamental design flaws, continued operation—and support for continued operation—entails a risk. Concluding that doing nothing is the greater risk, western governments stand with Weinberg, who notes, hopefully, that in his bargain with the devil, Faust won back his soul.

Chernobyl still casts a shadow over the entire post-Soviet nuclear sphere (see Figure 3.1). The accident continues to burden Belarus, Russia, and Ukraine with contaminated farms, forests, and waters, and Ukraine with a collapsing nuclear tomb. Ukraine claims that it has spent $14 billion on Chernobyl cleanup.[2] Thirty-one people, mainly firefighters, died as a direct result of the Chernobyl accident. Some 135,000 residents were permanently displaced. A population of 17 million people, mainly in Belarus and Ukraine, was exposed to radiation.[3] Evidence of chronic health effects is obscured by the quality of public health records and "increased

Figure 3.1   Chernobyl Unit 4, 1986
SOURCE: SPOT Satellite Corporation, © CNES 1999.

ascertainment," a statistical distortion explained by the fact that one finds more cancer cases when one aggressively searches for them.[4] Thyroid cancers have increased measurably in the region, with some 700 cases occurring since 1988 in the affected areas of Belarus, Russia, and Ukraine. An increase in leukemia, however, has not been detected, and some specialists are hopeful that the disease toll will not be as severe as expected.[5] This effect may stem from the relative immobility of cesium in soil caused by the fact that the element adheres to elements in the dirt. Nuclear advocates point to the low confirmable death toll as good news, arguing that Ukrainian coal mining each year kills more people than died as a direct result of the Chernobyl accident. On the other hand, nuclear opponents and some public health experts fear an epidemic of cancer cases in the making. Russian experts criticize western studies for having, in their opinion, seriously understated the health consequences of the accident.[6]

Chernobyl-style reactors continue to operate near St. Petersburg and Vilnius. These plants operate with the same risk as Chernobyl but get far less attention. Chernobyl may even be a lower safety priority because it is now surrounded by a 30 kilometer (18 mile) radius exclusion zone into which unauthorized persons may not go. Millions of people live within this range of the other operating RBMK (Chernobyl style) reactors.

A threat much graver than any nuclear reactor accident is posed by hundreds of tons of plutonium and highly enriched uranium stored in Russia with uncertain security. Nonproliferation may be the most important priority for international cooperation in the region because of the terrible destructive power of even crude nuclear weapons and because weapons-usable plutonium and highly enriched uranium appears to be vulnerable to theft. Only a relatively tiny amount, 7 kilograms[7], is required to make a weapon capable of devastating the center of Moscow, or most of Manhattan.

Policy makers grappling with nuclear power and weapons security must sort through complex technical and political problems. These matters require some understanding of nuclear science, engineering, and health physics. The critical subjects—how reactors work, how bombs are put together, how radiation affects the human body—are accessible to most people with a university education. Nuclear politics, on the other hand, remains as elusive as the neutrino.

Soviet Reactor Technology

The Soviet nuclear power industry built three basic reactor types: the RBMK, the VVER 440, and the VVER 1000. The RBMK, the Chernobyl-type reactor, was based on a pressure tube design used in the world's first nuclear power plant in Obninsk, not far from Moscow. The acronym RBMK is Russian and stands for large capacity, mixed tube reactor. The RBMK ranks among the world's largest power reactors, with 1,000 and 1,500 megawatt versions, the latter large enough to power a small city, say, the size of the District of Columbia. The "mixed" channels contain both nuclear fuel pellets, which are small, low-enriched uranium cylinders wrapped in a hard zirconium alloy steel, and the reactor cooling water. The channels, or tubes, are mounted vertically in a block of graphite, which serves as the moderator. The cooling water both protects the reactor from overheating and makes steam to produce electric power.

The acronym VVER stands for water-cooled, water-moderated power reactor. A VVER 440-230 is an older model with a power capacity of 440 megawatts. The newer model VVER 440-215 produces the same amount of power but is considered safer because it incorporates better passive containment, meaning that in the event of a catastrophic failure of safety systems, a containment vessel and other suppression systems would help to keep radiation from escaping the plant. Both the RBMK and the VVER model 230 lack adequate containment. The concrete reactor containment dome familiar to Americans and Europeans does not exist on these reactors. The VVER-1000, a later model and larger reactor, has containment comparable to that found in the west.

Twelve of fifteen currently operating RBMK power reactors are located at Kursk, Smolensk, and Saint Petersburg in Russia (see Figure 3.2). Two others are located in Lithuania, and one remains in operation at Chernobyl. Four small graphite reactors operate at Bilibino in the Russian far east. Six specialized reactors, also based on graphite, present even more serious safety concerns because they produce large amounts of plutonium. Three were built to make weapons material: one near Krasnoyarsk and two near Tomsk, both in Russia. A fast neutron reactor on the Caspian in Kazakhstan provides desalinated water.

Reactor Basics

Light water reactors, including the RBMK, work quite simply. Water is the working fluid, meaning that water is used to convert heat to power. Water is pumped over hot nuclear fuel, converted to steam in a single loop that goes to a turbine, or pumped under pressure to a heat exchanger where water in a second loop is heated, producing steam to turn a turbine to produce power. A single-loop reactor is called a boiling water reactor, whereas a water reactor with two loops is called a pressurized water reactor. Both types are called light water reactors because they use normal water, a molecule composed of two hydrogen atoms and one oxygen atom. Heavy water reactors use water composed of deuterium, a heavier isotope of hydrogen, and oxygen. French, Japanese, and U.S. manufacturers have produced commercial reactors of boiling and pressurized light water reactor types, whereas Canada has manufactured a commercial heavy water power reactor. The RBMK is a boiling light water reactor, having only a single loop of normal water running through the reactor core, boiling, turning the power turbine, being condensed back to water, and pumped back to the reactor core.

A nuclear chain reaction is not easy to start or sustain. Like Goldilocks' porridge or a market economy, it should not be too hot or too cold. Chain reaction requires that neutrons be absorbed in nuclei of fissile atoms such as uranium-235, and absorption occurs most often at moderate energy levels. When uranium atoms split and eject neutrons at high speeds, the neutrons must be slowed—moderated—before they can be absorbed by other fissile nuclei. When sufficient neutrons are available at moderate energy levels and sufficient fissile nuclei are arranged in a suitable geometry, a self-sustaining chain reaction, called criticality, is obtained. If the neutrons are not moderated—if they speed on at high energy levels— they can pass out of the reactor core into containment walls, and the

Figure 3.2   Soviet-Designed Nuclear Facilities.
SOURCE: U.S. Department of Energy, "Cooperation to Improve the
Safety of Soviet-Designed Nuclear Power Plants," (n.d.), 4.

chain reaction stops.[8] At energy levels too low, the neutrons do not have
the energy to penetrate the nuclei of fissile atoms, and the reaction stops.
The moderator can be water, as it is in all U.S. civilian and in all Russian
VVER-type reactors, or graphite, as in Russian RBMK and both Russian
and former U.S. plutonium production reactors.[9]

Control rods halt chain reactions by absorbing or over-moderating
neutrons, reducing availability of neutrons for fission to the point that
the reaction cannot be sustained. The core of an RBMK is made of 1,600
fuel channels, each holding about 115 kilograms[10] of enriched uranium
dioxide fuel. The fuel is enriched to about 2.4 percent concentration of
uranium-235, the fissile, or fissionable, isotope of uranium. The remainder is non-fissile U-238,[11] meaning that a 1,000 megawatt RBMK reactor
on average contains about 4 tons of fissile and 186 tons of fertile uranium.
The annual burn-up is a little less than 17 metric tons of uranium per

year. In contrast, a comparably sized coal-fired power plant consumes about 2.5 million tons of coal per year.

In the early 1970s U.S. experts assumed that, given any significant level of dependence on nuclear power, a reactor accident was inevitable. The question was whether it could be contained. This is why U.S. reactors were built with multiple layers of protection: extra electric pumps in case the main ones failed, diesel generators to power the pumps in case all power was lost, backup diesel generators, an emergency core cooling system if all the pumps failed, and a 2-meter-thick containment dome to hold material inside the reactor building in case of a serious accident. Three Mile Island proved the wisdom of that approach and the designer's philosophy that one should expect accidents and plan to contain them with multiple redundancies. Continued operation of RBMK and VVER 440-230 reactors, all of which lack containment vessels, has been described as a form of Russian roulette.[12] Some thirty Soviet-built power reactors lack containment vessels, including sixteen Chernobyl-type or RBMK reactors.[13]

Soviet reactors have other serious flaws. The RBMKs are over-moderated by their graphite and water configurations, which accounts for the infamous "positive void coefficient." The practical importance of this characteristic is that if an RBMK reactor loses cooling water, the reaction speeds up. That result contrasts with a western reactor which, if it loses water, slows or shuts down. The water in the western reactor is the moderator, and if it is lost, the neutrons speed up and no longer can sustain the chain reaction. But the RBMK still has graphite to moderate the reaction even if the water is lost. Even steam bubbles create voids that, in an RBMK, speed up the chain reaction and can result in the operator losing control of the reactor. Many experts consider that the positive void coefficient of RBMKs alone makes all reactors of that design inherently unsafe.

A related flaw stems from the design of RBMK control rods. The rate of reaction is controlled—and shut down—with control rods that can be inserted to capture neutrons and thus slow or halt the nuclear chain reaction. The RBMK reactor contains 211 absorber rods made of aluminum and containing hollow boron carbide pellets. Until recently, all RBMKs, including the ill-fated Chernobyl Unit 4, used control rods with graphite "displacers" at their tips. These displacers served to push water out of the bottom of the tubes that held the control rods when the rods were reinserted in the reactor core. Displacing the water with graphite created a condition analogous to formation of a steam bubble. Graphite is a more effective moderator than water, and substituting it for water accelerates the chain reaction in the immediate vicinity. So, instead of absorbing neutrons and slowing or stopping the nuclear chain reaction, the control rods initially, upon insertion, accelerate it. During normal operation, this effect is relatively unimportant, but if reactor conditions become unstable— such as during the experiment that precipitated the Chernobyl accident—inserting the rods can further destabilize the reactor. The RBMK was made less controllable by control rods that were gravity fed into the core. Several tens of seconds were required to insert the rods fully into the reactor. The reaction by that time was out of control, and a steam explosion "disassembled the reactor," as one report put it. Since the accident, pneumatic systems have been retrofitted to force the control rods fully into the reactor within three seconds of the operator's command. Also, the graphite displacer tips at the end of the control rods have been replaced with a non-moderating material.

The Chernobyl Accident

The great irony of the Chernobyl accident is that it was caused by a safety test. The test was designed to determine whether the reactor's power turbine could be tripped, disconnected from the external grid, and yet still provide enough power from its turbine to operate pumps to feed cooling water into the reactor. Reactors, once they are shut down, generate enough heat from radioactive decay products to melt the fuel core unless external cooling is provided. The idea was to shut down the reactor and test the ability of the turbine to provide power of sufficient voltage to keep the pumps working long enough to cool the reactor. The turbine, after all, continues to spin for a time after power is lost, and can continue to receive steam from the residual heat and radioactive decay in the reactor core.

The operating log of the Chernobyl reactor tells the story of the mistakes that led to the explosion. Re-reading it reveals at least three flaws in the system contributing to the accident. These shortcomings include the instability of the reactor (positive void coefficient), lack of containment, and lack of a safety culture that would have prevented many of the actions taken by operators during the test.

The objective of the experiment was to test the performance of the turbo generator after the reactor was tripped. The generator was disconnected from the grid, meaning that it was no longer supplying power to the outside world. To remove other variables from the test of the turbo generator—that is, to remove the chance for water to be supplied to the reactor by other means—the emergency core cooling system was manually shut off. Auxiliary power equipment not involved in the test was disconnected.

The reactor was thus put into a dangerous state that reduced water flows and then restored them suddenly. This condition, coupled with the increase in power, caused boiling in the bottom of the reactor. With steam

---

BOX 3.1   EVENTS LEADING TO
THE CHERNOBYL ACCIDENT

The test performed on the RBMK at Chernobyl Unit 4 began on 25 April 1986. It was intended to determine whether sufficient reactor cooling water could be supplied from pumps powered only by the reactor's turbo generators–after the reactor had been tripped or shut down. The safety test began with a reduction of power. By *03:47*, thermal power was at 1,600 MW, half the maximum. The reactor was generating 720 MW of heat at *00:05* on 26 April. Thermal power dropped to only 30 MW, below the planned level. Operators raised power, and by *00:34*, steam levels exceeded levels that would have engaged emergency systems, had they not been shut off. *00:41 to 01:16*: The test continued with second turbo generator disconnected from grid to test for vibrations while operating without load. The emergency system that would have shut down the generators was disabled. Reactor heat reached 200 MW by *01:03*. Water pumps were switched on at *01:03*, Water flowed unevenly, temperatures increased, an alarm sounded at *01:18:52*. Turbine test began at *01:23:05* when the turbo generator's valves were closed. At *01:23:40*, operators pressed an emergency button, control rods were inserted. Alarms signaling a reactor runaway sounded at *01:23:43*. At *01:23:47*, coolant flow dropped by 40 percent, steam pressure rose, and fault signals sounded. At *01:23:48* water flow rates were restored, but steam levels kept building, and steam dumping and condensing safety systems for the turbines were activated. At *01:23:49* an alarm signaled a fuel channel rupture and faults in the control rod mechanisms. "Severe shocks" were recorded at *01:24* and the control rods stopped, and their power shut off. The reactor was blown apart by a steam explosion when control rods were inserted.

---

formation, the "positive void coefficient" inherent in the RBMK accelerated the nuclear reaction, which produced more steam bubbles, which further sped the reaction, and so forth. Water pumps began cavitating, failing to pump water, so that even water entering the reactor was nearly boiling. The emergency core cooling system was unavailable because it had been disabled by the operators to conduct the safety test.

An area of instability formed near the bottom of the reactor core where water had boiled to steam; the reaction rate was already dangerously high. The reactor operators, realizing at long last that they were losing control of the reactor, scrammed it by releasing the gravity-fed control rods into the core. When the graphite displacers at the end of the control rods reached the bottom of the core, they had the effect of accelerating the reaction rate. Insertion of the rods, meant to stop the chain reaction, accelerated it, causing a steam explosion and shock wave that destroyed the reactor and started an uncontrollable graphite fire.

An operator on duty at Chernobyl the night of the accident described the experience: "We heard no explosion, but felt some moving." He and his colleagues believed that the steam turbine, spinning rapidly under great pressure, had come apart. The operator ran to the turbine room of Unit 4 and opened the door. "I was shocked to see that the walls of the turbine room were blown *in*," he said. If the turbine had experienced a steam explosion or if the turbine had blown apart, the walls would have been blown out. He encountered a colleague who had been near the explosion, who received an estimated 800 REM, and who was the first to die from the accident.[14] (See "health effects" in Box 3.2). The young operator, who told his story in 1996 from the control room of then-operating Chernobyl Unit 1, received an estimated 80 REM in the accident. He said that now he felt fine and ran his hand through his bright red hair, joking that it "had not fallen out." He mentioned that possibly his blood was "not quite right," but that his health and life were otherwise normal.

Chernobyl today, of course, is not normal. The fields around the plant and row upon row of streets in the nearby town are empty, quiet, overgrown. Journalists and politicians describe the eerie sensation of viewing empty flats and a rusting Ferris wheel. More than 100,000 people who lived in the city have been moved to newly built villages in Kyiv oblast. A quarter of them now live in Slavutych, a new city 50 kilometers from the plant, built for plant workers. A fence marks the perimeter of the area—it has a radius of 30 km—in which no one is permitted to live and into which the general public is not allowed. Deep forests line both sides of the road. A sign in three-foot-high Cyrillic letters says the forest is "good for the health," but another large road sign warns, "Radiation doses are higher in the forest."

The small city of Slavutych, built after the accident to house power plant workers and their families, has become a focus of the struggle to close down the power plant. "This is a one-industry town," Slavutych Mayor Vladimir Udovichenko said in 1998. "Even the babies work for the nuclear power plant."[15] The effort to close Chernobyl has been an exercise in frustration. Ukrainian officials complained of broken western promises to provide billions of dollars to complete VVER reactors to "compensate" Ukraine for closing two operable units at Chernobyl.[16] Western leaders—privately—accused Ukraine of blackmail, of holding hostage for money the safety of their citizens and neighbors.

The G-7 did promise assistance in return for closure of Chernobyl, but the money was offered in the form of loans, which Ukraine would have to pay back. The funds, to be administered by the EBRD, would finance com-

BOX 3.2   RADIATION CONCEPTS

The Three Mile Island reactor accident released 20 curies of different types of radiation; the Chernobyl accident released 50 million curies of cesium and strontium, probably more short-lived radioactive iodine. How does either relate to health effects? A visitor to Chernobyl city in 1998 received 0.35 millirem per hour, and got a total of 50 millirem touring the sarcophagus. Will she be ill?

The curie, as a unit of measure, tells little about biological effects. One curie is simply the amount of a radioactive material that will produce 37 billion decays–37 billion emissions of alphas, betas, or gammas–per second. This happens to be the number of decays per second in one gram of radium, the element studied by Pierre and Marie Curie. Similarly, the roentgen is a measure of ionization potential in air. This measure provides more information about the energy produced by radiation. The RAD, a related measure, is an acronym for radiation absorption dose, and it equals 0.01 joule per kilogram (about 0.01 BTU per ton). It is a measure of the energy absorbed per unit mass of material, such as air or shielding.

A useful measure of radiation dose in humans is the "roentgen equivalent man," or REM. This value is adjusted for the amount of energy effectively delivered to human tissue. Radiation health effects depend on three factors. First is the type of radiation. Alpha, beta, and gamma radiation penetrate tissue at different rates and deposit different levels of energy. An alpha particle is a bundle of two protons and two neutrons. A beta is similar in mass and charge to an electron. An alpha particle is 7,200 times more massive than an electron or beta particle, and thus is very likely to collide with atoms in flesh. As a result, it far less likely to penetrate skin, or even a thin sheet of paper. Consequently, the alpha particle will present less of a health threat unless the alpha source is ingested or inhaled, where radiation can deposit energy in softer, more vulnerable flesh. Tearing through lung tissue, for example, an alpha particle would tear through flesh, ionizing and destroying cell structures. Beta particles penetrate more deeply and can cause similar damage. Gamma rays may pass through flesh missing everything. Or one might strike a molecule of DNA in a cell nucleus and break its molecular bonds. If the DNA strands were to reform in sequences different from healthy genes, mutations, and sometimes cancer, can result.

A related concept is relative biological effectiveness. It is a multiplier for the effect that biological concentration, for example, would have on radiological dosage. For example, radioactive strontium can be concentrated in cow's milk when cow's graze on contaminated grass because strontium mimics the chemical behavior of calcium and displaces it in cow's milk.

(continues)

(continued)

The REM in humans provides a useful framework for relating natural background radiation–from the radioactive materials in stone, concrete blocks, from cosmic sources–to medical and nuclear industry exposures. Most Americans and Europeans receive fewer than 200 millirem (0.2 REM) each year from background. A chest X-ray delivers roughly 50 millirem (0.05 REM), and a mammogram one full REM. Radiation sickness occurs at 100–150 REM, usually over a 2 week period. An important measure is the LD–50/30–the lethal dose for 50 percent of the population in 30 days. The commonly accepted value is about 500 REM. Radiation afflicts more than 260 areas in Russia, 40 of which had radioactivity levels greater than one milliroentgen per hour.

Radiation Health Effects

Item	Exposure (REM)
Chest X-Ray	0.05
Natural, per year	0.2
Annual occupational limit	5.0
Acute Radiation Sickness	100.0
LD-50/30	500.0

NOTE: International convention for radiation units uses the Gray (Gy) in lieu of the RAD and the Sievert (Sv) in lieu of the REM. One Gray equals 100 RAD, and 1 Sievert equals 100 REM. The older units remain in use.

SOURCE: Carl R. Nave and Brenda C. Nave, *Physics for the Health Sciences* (Philadelphia: W. B. Saunders, 1980), 341–346.

---

pletion of "replacement" reactors, two VVER-1000 reactors that were substantially complete. The loans, among other things, would pay for safety upgrades to be made by western companies. This scheme was flawed, however, because EBRD's rules require that loans be made for cost-effective investments, which in the case of Chernobyl meant that the VVER-1000s had to be a cost-effective means of meeting electric power demand.[17] An earlier assessment by EBRD consultants had claimed to have proven that completing the reactors was cost-effective. A subsequent EBRD study of the $1.8 billion cost of completing Khmelnitsky Unit 2 and Rovno Unit 4—both 70 percent complete—characterized the project as not cost-effective and threw in doubt the G-7's ability to keep its promises.[18] EBRD subsequently postponed, then revived, then again reconsidered its decision on the replacement reactor financing.[19] The G-7, however, committed funding to help stabilize the sarcophagus entombing Chernobyl Unit 4, and Ukraine finally agreed to close the plant in 2000.[20]

If capacity were the only consideration, Chernobyl could easily have been closed at any time during the 1990s. Ukrainian base load power demand collapsed with the economy, and peak power demand rarely exceeded 50 percent of installed capacity. The nation had roughly twelve times more fossil power capacity sitting idle than in use at Chernobyl. However, instead of importing Russian gas or oil to generate electricity at idle fossil fuel plants, Ukraine used nuclear power from Chernobyl as a fuel saver, although Chernobyl's power, at most, represented only about 1 percent of the total energy imported from Russia. A problem meeting peak demand grew severe in 1998 as nonpayments for electricity increasingly meant that the fossil fuel was unavailable to meet power demand. The nation had great difficulty paying for energy imports and concluded that the risk of repeating the Chernobyl disaster was the lesser of two evils. Moreover, Ukraine has had difficulty even paying Russia for the relatively small imported fuel cost for its reactors.[21]

Independence and statehood somehow transformed Chernobyl from a bitter symbol of Soviet imperialism and environmental mismanagement into an emblem of Ukrainian national sovereignty and survival. The change of spelling from Russian to Ukrainian is the least manifestation of this change. The Ukrainian government's insistence on western compensation for shutting Chernobyl—an act that would protect its own people more than anyone else—characterized that system in a way Faust could appreciate.

Ukrainian officials have consistently maintained that closure of unsafe reactors was contingent upon the replacement of that lost capacity with other nuclear capacity. A World Bank/EBRD mission concluded that such subsidies would make sense only if an overall agreement for the management of nuclear systems were agreed upon between the G-7 and the Ukrainian government[22] One Chernobyl shutdown plan required EBRD debt financing well beyond the grant making capacity of its Nuclear Safety Account, and Ukraine negotiated hard to obtain funds from the G-7 to complete five Soviet-designed reactors that were left incomplete after the breakup of the Soviet Union. These included a reactor at Zaporozhe that was 95 percent complete, one each at Khmelnitsky and Rovno, both 70 percent complete, and two others at Khmelnitsky that were less than one-fifth complete.[23] The Ukrainian State Committee for the Utilization of Nuclear Power (Goskomatom) hoped to raise $2.1 billion to complete the five, but the EBRD disputed the cost estimate. The G-7 experts put the cost of completing just the two reactors that were 70 percent complete at $1.5 billion, implying that the total cost for all five could be many times that sum. One difference in the estimates was the level of western safety equipment assumed to be purchased and installed.

The scope of problems at Chernobyl goes far beyond closing it. The sarcophagus, or shelter as plant managers prefer to call it, must be stabilized forever on human time scales, and doing so would be a challenge even for wealthy western nations with extensive nuclear clean-up experience. The sarcophagus—the concrete shelter or building cobbled together—is unstable due to the weight of concrete on the foundation, which was not built to support it. The G-7 countries have offered $300 million to rebuild the shelter, an amount unlikely to solve the problem.[24] Even at that price, however, the opportunity cost of spending so much to accomplish so little, as has been the case at U.S. nuclear waste sites such as Hanford, Washington, may compel authorities to consider postponing clean-up and instead resort to "dogging and fencing" the site, as one observer put it.

Aiding Nuclear Power

Beyond Chernobyl, western policy toward post-Soviet commercial nuclear power is driven by at least three major concerns. One is security.[25] As long as reactor fuel and weapons are linked and nuclear workers are paid poorly and late, vigilance must be maintained against nuclear proliferation and terror. Another priority concern is safety, because the human and economic cost of "another Chernobyl" would be high. Not only would the lives of many persons in the region be devastated and their property destroyed, but the west itself might incur major costs in aiding a clean-up and mitigation response. Another concern is commercial. Western vendors of reactor components and controls see the region as a source of orders in an otherwise depressed global market for nuclear power. Related to this interest is western fear of the impact an accident would have on commercial nuclear industry. That is, the western nuclear industry believes that an accident at an unsafe Russian reactor might cause a public outcry against nuclear technology more generally and lead to its own demise.

Energy availability is a more distant concern. Nuclear power dependence varies across the region, but with important exceptions, remains generally low. Lithuania obtains four-fifths of its power from four RBMK reactors.[26] Ukraine produces about 45 percent of its electricity in nuclear plants, having shut down much of its fossil fuel capacity to reduce gas and oil imports. Other nations producing roughly two-fifths of total power supply using nuclear plants include Bulgaria, Hungary, the Slovak Republic, and Slovenia. The Czech Republic obtains about one-fifth of its power from nuclear sources. These nations, with the exception of

Ukraine, are all quite small, and larger countries rely much less heavily on atomic power. Russia gets less than 12 percent, Kazakhstan less than 1 percent, and Poland and Romania no power from nuclear sources. In comparison, nuclear power provides about one-fifth of U.S. electric power. The total energy—as opposed to power-generating capacity—produced by nuclear stations in the entire former Soviet-Eastern European region amounts to only 1.5 exajoules, a little less than 2 percent of total primary energy use.[27]

Foreign nuclear assistance otherwise relates mainly to the safe—but continuing—operation of inherently unsafe reactors, including all the RBMKs and the VVER 440-230s. Russia in the 1990s operated eleven RBMKs with a power capacity of about 11 gigawatts. These reactors could produce a maximum of 0.7 exajoules of electricity per year, worth about $2.5 billion annually. Several older, less safe versions of the VVER design also continue to operate. Two VVERs in Armenia, four in Bulgaria, four in Russia, and two in Slovakia remain operable or in operation. A total of twelve RBMKs and older VVERs with a capacity just under 10 gigawatts now operate outside of Ukraine and Russia.[28] That includes Lithuania's Ignalina RBMK station Bulgaria's four VVER 440-230s, and Slovakia's two VVER 440-230s. Armenia also possesses two operable VVER 440-230s, both of which were shut down during Soviet times because they were located near severe geological faults. Authorities then considered the reactors unsafe for operation regardless of their design and maintenance. However, Armenia, squeezed by war and an Azeri energy embargo, have pressed one reactor back into service despite the risk, although some risk mitigation has been done.

Both safety and security are threatened by the dire circumstances of the region's nuclear enterprise.[29] One of the key safety issues today in the region is the ability to maintain a qualified work force. Lithuania was an extreme case in that the operators for its two RBMKs were Russian citizens who might have left the country after independence. The government of Lithuania addressed the problem by giving citizenship to the operators, although it has more recently indicated willingness to close the plants by 2005.[30] Elsewhere the problem is money. One million people dependent on the Russian Ministry of Atomic Energy (Minatom)—a figure that includes not only nuclear workers but maintenance staff, teachers, doctors, and all the persons who support the nuclear cities and their facilities—get paid late, partly because payments for power generation are sporadic. Nuclear power plant operators have had to strike to get paid. Operators do not make as much as they might expect given the level of responsibility they carry, and worse, they often do not get paid at all. Anger over not being paid prompted a dozen employees at the Leningrad Nuclear Power Station near St. Petersburg to seize the control room and threaten to shut down the plant. When they were joined by 400 colleagues and declared a hunger strike, the federal government flew cash in to pay the workers. A spokesman for Minatom said, "They are right to protest."[31] Atomic workers at the Smolensk plant took a different approach, making a protest walk some 360 kilometers (approximately 215 miles) to Moscow. They, too, were successful, meeting with the first deputy prime minister and obtaining a promise of payment.[32] Nuclear safety in Ukraine suffered from the loss of qualified personnel, some of whom left for Russia where they could earn more money. In Ukraine, salaries were limited by regulations and by the fact that over four-fifths of nuclear-generated power was not paid for by customers.[33]

Technical Assistance

At Chernobyl, international technical assistance has been directed at short-term safety efforts. Fire prevention—adding fire doors, replacing flammable floor coverings—has been a focus, as has operator training. These actions accepted that the plant would continue to operate. The U.S. government supported an energy efficiency program designed to help reduce energy imports by financing energy efficiency upgrades.[34] Stabilizing the Unit 4 sarcophagus has become a major focus, motivated by European fears of being exposed to radioactive dust from the shelter's collapse.

Unlike the World Bank, which has steered clear of nuclear power, EBRD has taken on nuclear safety subsidies as a major part of its portfolio. The EBRD has created a special nuclear account with roughly $300 million that EBRD describes as necessary to give priority to "those reactors presenting a high level of risk that can be significantly reduced by short-term safety improvements, and that are necessary to ensure the continuing national electricity supply in the region."[35] EBRD used the funds in Kozloduy[36], Ignalina, Chernobyl, and three Russian plants.[37] The funds were provided in the form of grants, which the EBRD handled for the G-7 donor nations. Intense controversy has developed around the account, as well as G-7 policy. Client nations, particularly Ukraine, have bitterly complained that the west has not kept its promise to provide compensation for closure of the reactors. Economists and safety experts have complained that, in Ukraine in particular, the plan to provide money to complete new reactors in return for closing the unsafe ones does not meet EBRD's own rules requiring cost-effectiveness. Environmentalists have complained that the G-7 erred by encouraging Ukraine and Slovakia to complete new reactors, and have accused them of promoting their own nuclear industries. The nuclear industry, mainly Siemens, Westinghouse, and Framatom, complain that the west made a mistake by insisting on closure rather than working on safety upgrades, and that the east has not competently installed regulatory and market infrastructures necessary to manage nuclear power. The client countries, especially the Czech Republic, also have complained that the western companies keep failing to meet deadlines for installation of control and safety equipment.[38]

Environmentalists have strongly criticized the G-7 approach to closing Chernobyl, comparing it to the mistake of building Chernobyl in the first place.[39] Their criticism extends to nuclear cooperation policy across the region. Environmentalists have long and sharply criticized G-7 and "Big-8" nuclear policy, and especially after a "Nuclear Safety Summit" in Moscow in 1996 involving heads of state, including presidents Clinton and Yeltsin. Environmentalists had hoped the Big-8 summit would achieve the following:[40]

- Identify the most dangerous reactors to be shut down and provide for financing for replacement power and decommissioning.
- Create a civilian regulatory and safety inspection framework, and provide for public participation in the oversight process.
- Develop a safeguards regime for and public oversight of fissile inventories.
- Halt plutonium reprocessing, end the use of highly enriched uranium, and assist the civilian conversion of Russian plutonium economies at Ozersk, Seversk, and Zheleznogorsk.[41]

An "alternative" summit of environmentalists harshly criticized the outcome of the summit, accusing the Big-8 leaders not only of failing to reduce nuclear dangers but even of increasing the potential for nuclear accidents and proliferation. They noted that existing nuclear plants remain "competitive" because fuel costs are minimal and capital costs are amortized or were borne by government. At the margin, they suggest, nuclear power is probably not competitive with gas.

The Natural Resources Defense Council (NRDC) concluded that, "Dangerous reactors with inherent, uncorrectable design flaws are to be kept on line with Western assistance for an indefinite period."[42] Indigenous activism has also been on the increase. Citizens in Krasnoyarsk demanded a referendum on halting a nuclear fuel reprocessing facility, where some 100,000 persons signed a petition calling for the vote.[43]

Minatom's problem is that the future is bleak unless customers order new plants. That explains the aggressiveness with which former Minatom chairman Victor Mikhailov fought to keep operating stations going or to replace them with new or retooled nuclear plants. Minatom's desire for self-preservation has put western governments in the position of sacrificing concern for safety in return for achieving security goals. That was the case when the United States capitulated on its demand that the Seversk and Zheleznogorsk plutonium production reactors be closed and replaced with non-nuclear plants. Instead, the reactors were to be reconfigured, in part with U.S. funding, to produce much less plutonium. This plan, too, stalled.

Minatom has aggressively marketed its systems abroad, underbidding western nuclear power vendors for reactor completions in Slovakia and rebuffing all U.S. efforts to block its reactor sales to Iran. In addition to reactor deals with China and India, Minatom secured business in Slovakia by agreeing to provide an $80 million loan to Slovakia for completing a nuclear plant at Mochovce. To be approved, the loan had to be, and was, included in the Russian 1997 state budget. The Austrian government earlier had offered a $50 million contribution for alternatives if Slovakia would not build the plant.

Minatom's ability to frustrate western negotiators stems from three advantages. First, foreign governments are dependent on Minatom's good will even to have access to nuclear installations. Second, U.S. companies want Russian customers for reactor upgrades, so they have to be polite. Third, Minatom is a complete monopoly and one able to generate political strength with nationalist and patriotic pride. As a government-owned enterprise, one not driven by market forces, Minatom has been able to prop up a commercially unsustainable system.[44] That is, state intervention has helped sustain an uneconomic system.

Eastern European Reactor Deals

Eastern Europe's nuclear industry is less desperate because it is smaller. As in Russia, the key issues are jobs, safety, cost, energy independence, and business.[45] Politics shapes the debate across the region more than economics because nuclear power remains a state-owned enterprise.

A former commissioner of the Nuclear Regulatory Commission, Ivan Selin, suggested that making the region's reactors safe would require $20 billion in financing. Financing requires economic viability and acceptable risk. Guarantees would have to be provided so that investors would not be liable for damage from any accidents and so that they would recover their investments even if the investments were destroyed.[46] Financial risk is a main reason no utility has ordered a nuclear power plant in the United States since the 1970s: No investor wants to lose a billion dollars overnight as did the investors at Three-Mile Island.[47] The question is whether nuclear power is worth massive public subsidy.

Many of the former centrally planned economies are seeking foreign credit to rehabilitate unfinished or substandard power stations. A case study in this approach is the U.S. effort to complete construction of Temelin, the Czech nuclear power station in Southern Bohemia. CEZ, the Czech utility, convinced the Czech and U.S. governments that the facility was key to meeting future energy demand. In 1992, Westinghouse Electric Corporation signed a contract to rehabilitate two VVER-1000 reactors in a four-reactor nuclear complex at the Temelin plant in Southern Bohemia, Czech Republic. Construction of the power station had been halted after the 1986 Chernobyl disaster.[48] Westinghouse agreed to supply technology, equipment, and services, including new control systems and nuclear fuel for each 1000-megawatt reactor, in a deal valued at $350–400 million.[49] A crucial element of the contract was a loan guarantee provided by the U.S. government through the U.S. Export-Import Bank.

Opponents of the venture include the government of Austria, Greenpeace, the National Resources Defense Council, Friends of the Earth, and other environmental groups The Czech ecological movement *Hnuti Duha* (Rainbow) collected several hundred signatures from residents in southern Bohemia to protest completion of the plant, blocked entrances to it, and set up a protest camp nearby.[50] Opponents cited an internal World Bank study contending that the Czech Republic would have no need for Temelin's power until 2010 and favoring a gas-fired system for eventual replacement. The Czech government dismissed this conclusion.[51] Independent analysts had also suggested that improved energy efficiency would obviate need for the plant. The World Bank even offered $250 million in credits for energy conservation and environmental controls at the lignite-fired plants, but this loan was stopped when former prime minister Vaclav Klaus—for other reasons—invited the World Bank to leave the country.

Advocates of the project observed that the Czech Republic was heavily dependent on coal for power generation, with over 12 of 19 gigawatts in the country provided by coal-fired thermal plants burning mostly lignite with low caloric value and high sulfur content.[52] Winter air pollution inversions in Northern Bohemia are greatly exacerbated by quantities of sulfur dioxide, ash, and particulates emanating from smokestacks. The Temelin project was supposed to replace roughly 16 percent of the power generated by coal-fired plants.

The role of western governments remains the most interesting aspect of this deal. Some high-ranking Czech officials quietly complained that the U.S. government exerted pressure on the Czech government to continue operation of the plant and to ensure that Westinghouse be selected to upgrade it. An internal Ministry of Foreign Affairs document reported the following communication after a meeting of officials of the two governments:

The U.S. government would be greatly displeased by the termination of the Temelin reactor. .... The U.S. government would be greatly pleased by the selection of Westinghouse to upgrade the safety of Temelin.[53]

These sentiments were communicated in a meeting in Washington with high-level officials in late winter 1993. When asked about this criticism, some officials suggested that the democratic government of a hero such as Vaclav Havel, president and playwright, should have been able to make its own decisions about the use of nuclear power. Whether the U.S. government should be promoting and subsidizing energy use is a separate question. Temelin raises other questions about the advantages of nuclear power generation weighed against environmental concerns and alternative technological options. Are risks from air pollution comparable to risks from nuclear energy? Is there a cheaper way to meet the region's energy needs? However these questions are answered, construction delays have pushed full completion of Temelin back to the year 2005 and have greatly increased its costs.

Similar questions arose regarding a plant at Mochovce in the Slovak Republic. Mochovce is a Russian-designed pressurized water nuclear power plant in the VVER 440-213 series. It had not been completed at the time of the breakup of Czechoslovakia, and western governments had hoped to upgrade the plant with modern control equipment and use its completion as a bargaining chip with the Slovak government.[54] As in Ukraine, where the west offered to finance completion of new reactors in return for closing Chernobyl, the EBRD offered to finance Mochovce in return for closure of a VVER 440-230 station. However, when a consortium led by Electricité de France applied early in the 1990s for a loan from EBRD, EBRD conditioned approval of the loan on the results of an analysis to determine whether Mochovce would be the least-cost alternative. Meanwhile, the Slovakian decision to push forward with Mochovce created tension between Slovakia and bordering Austria, which has imposed a nuclear moratorium in its own territory.[55]

A consultant to the EBRD at first concluded that completion of the plant was, in fact, the least-cost means of providing the power. Critics, however, identified shortcomings in the analysis, which cast additional doubt on the conclusion. For example, future energy demand—and the need for the Mochovce plant—was estimated using a simple econometric analysis. That analysis made unusual assumptions, including a price elasticity of demand for electricity of –0.16. An assumption taken from the economic mainstream would have been much higher, on the order of –0.7 to –1.1.[56] The higher assumption would have obviated the need for new sources of power. Modeling regional demand properly, of course, is a general concern that goes beyond nuclear power.

The G-7 similarly sought at Kozloduy, Bulgaria, to obtain closure of two VVER 440-230s, again by helping to finance modernization at other reactors, this time two other units at the same station. The EBRD had obtained a commitment from the Bulgarian government to close units 1 and 2 at the plant by the end of 1998 in return for a $26 million loan for rehabilitating four older units, including units 1 and 2. The deal ran into the same kind of problems as had similar arrangements in Ukraine and Slovakia, with the Bulgarian State Energy Committee asserting that the older plants would have to remain in operation until the new units were made available.

A major goal of western nuclear safety assistance has been to create a safety culture. The tools to this end have been grants and financial inducements. The incentives were meant to suborn behavior to close or upgrade plants considered a threat to the region's health and safety. Ironically, the aid has provided both the rationale and the tools to help keep inherently unsafe plants—plants with design flaws and without adequate containment—in operation. By helping keep open inherently unsafe reactors, the safety program undermines its own goal. No authentic safety culture would continue operation of reactors that could be readily replaced.

Civilian Nuclear Security and the Military

Civilian nuclear issues are complicated by increasing linkage with military nuclear operations, especially reductions in nuclear warheads. The gravest post-Cold War threat to global security is the risk of theft and diversion of nuclear weapons-grade material.[57] Harvard security analyst Graham Allison and his colleagues have documented six attempts to divert bomb-grade material. For example, Leonid Smirnov, an engineer at the Luch research and production facility in Podolsk near Moscow, was caught in 1992 at a railway station on his way to find a buyer for 1.5 kilograms (3.3 pounds) of highly enriched uranium. He had worked in the plant for 24 years and had smuggled out the material 50 grams (1.8 ounces) at a time.[58]

Four categories of Russian nuclear facilities are vulnerable to theft and diversion of weapons-grade material.[59] First, the Russian Ministry of Defense is responsible for safeguarding 45,000 nuclear weapons in its military arsenals. The Russian government set a goal of reducing its arsenal to 7,000 weapons by the year 2000, and during the 1990s it dismantled 3,000 bombs each year. Rapid progress toward this goal represents a major success story for the U.S. and Russian governments. The security of these weapons is reportedly high, and claims to the contrary are considered by some experts to be exaggerations.

Second, the assembly and disassembly of nuclear weapons involves significant risk of diversion. Minatom has responsibility for assembling and disassembling weapons. This work is conducted in some of the closed "plutonium cities." Weapons-grade plutonium and uranium are stockpiled as weapons and are disassembled in other facilities such as Tomsk-7 and Chelyabinsk-65 (also called Seversk and Ozersk, respectively) (see Table 3.1 and Figure 3.3). Dismantling 3,000 weapons each year adds an estimated 15 tons of plutonium and 45 tons of highly enriched uranium to the stockpile. This material is presumably secure for the present, but represents a large future threat. Third, the production and separation of reactor-grade plutonium, managed by Minatom, has created a stockpile of 30 tons of plutonium. This also presents a potential major long-term threat.[60]

Fourth, an archipelago of naval and research reactors has been built across the former Soviet Union. These facilities have been the target of the most serious known cases of attempted theft and diversion. For example, navy Captain Alexei Tikhomirov in 1993 broke a padlock on an unguarded submarine fuel warehouse in the Sevmorput navy yard in Murmansk and stole 4.5 kilograms (10 pounds) of 20 percent enriched uranium. Caught because he asked for help in selling the material, he wanted only $50,000. A year later, a Czech scientist was arrested in connection with 2.5 kilograms (6 pounds) of highly enriched uranium found in a car in Prague. The material was near-weapons-grade at 88 percent enriched.[61]

Deteriorating financial conditions in both the civilian and military sectors have complicated security. The Seversk and Zheleznogorsk plutonium production reactors, for example, were built to make plutonium for bombs. These were similar to the Hanford plutonium production reactors near Richland, Washington. With the arms race over, the security of all sides would be enhanced by closing the reactors and ceasing plutonium production. However, the reactors provided district heat to civilian populations in both cities and could not be shut off without replacing the heat sources.[62] An ironic joke stemmed from the fact that reactors formerly producing heat as a by-product of plutonium production now produced plutonium as a by-product of heat production. Unfortunately, the production rate of three operating reactors totaled 1.0–1.5 tons of plutonium per year, enough to make more than twenty-five atomic bombs. Russia and the United States negotiated cooperative measures to end this dangerous production, but, in another ironic twist, fossil fuel could not readily be supplied as a replacement despite the fact that Russia produces more natural gas than any other country, more oil than any country except Saudi Arabia, and more coal than any country except China and the United States. The Russian Ministry of

TABLE 3.1 Selected Russian Nuclear Weapons Facilities and Their Risks

Facility[a]	Function	Risk
Ozersk: Mayak Chemical Combine (Chelyabinsk-65)	Formerly plutonium production; currently fuel reprocessing	Diversion; plutonium production and release; waste dumping/leakage
Seversk: Siberian Chemical Combine (Tomsk-7)	Plutonium production; uranium enrichment; UO_x production	Diversion; plutonium production and release; waste dumping/leakage
Zheleznogorsk: Mining-Chemical Combine (Krasnoyarsk-26)	Plutonium production	Diversion; radiation releases
Sarov: Avangard Electro-mechanical Plant (Arzamas-16)	Weapons assembly, disassembly	Diversion, contamination

NOTE: a) Italicized names designate cities housing facility workers; parentheses contain Soviet era code names of the weapons facilities.

SOURCE: T. Cochran, R. Norris, and O. Bukharin, *Making the Russian Bomb* (Boulder: Westview Press, 1995), 40, 71, 137, 149.

Figure 3.3   Closed Nuclear Cities
SOURCE: U.S. Department of Energy.

Atomic energy preferred a nuclear solution and sought U.S. funds to re-engineer the reactors to keep producing heat while creating less pluto-nium. This effort involved "re-coring", or reconfiguring, fuel elements and moderator to greatly reduce the rate of plutonium produced in the reactor. The U.S. Department of Energy, early in the Clinton Adminis-tration, initially opposed this solution, but agreed to help finance the conversion of two units. Then-Secretary of Energy Hazel O'Leary and then-Minister of Atomic Energy (Minatom) Viktor Mikhailov signed an agreement in Washington in summer 1994 committing U.S. support to replace heat and power supplied by the reactors to the civilian popula-tion. The United States would have achieved its goal of reducing pluto-nium production, and Minatom would have achieved its goal of keep-ing revenues flowing and jobs in place for its workers. By January 2000, however, this option had become too costly and the two sides reopened negotiations to replace the reactors with fossil fuel-fired plants. Pluto-nium production continues, and local populations are placed at risk by continued operation of inherently flawed reactors. All three reactors are similar to the RBMK design. Although the Zheleznogorsk reactor is built inside a mountain and not likely to menace many people, the Sev-ersk reactors lack containment.[63]

A more global threat involves weapons-usable material generated by civilian reactors. Light water civilian reactors, like breeder reactors, produce plutonium, although unlike breeders they produce less energy in the form of plutonium fuel than they consume in the form of uranium.[64] Separating plutonium from spent civilian reactor fuel is difficult due to intense radiation. Minatom has already separated 30 tons of plutonium from used civilian reactor fuel, enough to make several thousand bombs. This work is done at Ozersk.[65]

Chelyabinsk-65's principal source of income is reprocessing spent fuel from VVER-440 reactors exported to Bulgaria, Finland, Hungary, and Ukraine. The complex separates 1–2 tons of reactor-grade plutonium each year. Seversk, with a population over 100,000, also produces about a ton of plutonium per year at the Siberian Chemical Combine, which also makes uranium oxide metal for blending to produce reactor-grade fuel.[66] The plutonium processing plant operates much like the "Purex" process formerly used in the United States at Hanford. This chemical process applies nitric acid, phosphate, and hydrocarbon solvents to separate uranium and fission products such as cesium and strontium from plutonium. In 1993, an explosion at the reprocessing plant at Seversk resulted in beta radiation contamination. Radionuclides with a half-life of one to two months contaminated at least two villages, causing health problems. Minatom concealed this fact until the Social-Ecological Union, a Russian nongovernmental organization with many high-level scientists as members, invited experts from Krasnoyarsk to perform measurements. After a hearing in the Tomsk oblast administration, a decision was taken to remove children from the villages, and former Minatom head Mikhailov acknowledged the exposures at a press conference.[67]

Seversk, or Tomsk-7, is on the Tom River, which flows into the Ob River. Only 15 kilometers (9 miles) from the Tom, millions of cubic meters of liquid radioactive waste were stored in a sandy underground layer 300–600 meters deep. Radiation from this waste will, it is feared, reach the Tom and then the Ob. This situation could be compared with that of the Hanford Reservation, where the United States produced plutonium during World War II and the Cold War, and where stored and leaking waste threatens the Columbia River. Unlike Hanford, however, Tomsk-7 is still producing and processing plutonium and uranium. Tomsk-7 is also not as well isolated from water supplies. Hanford's unshielded, graphite-moderated reactors have been closed; most of the leaked waste is suspended in soil and migrates very slowly toward the Columbia River. Significantly, the Siberian Chemical Combine as recently as 1997 lacked modern equipment for safeguarding nuclear material. Funds from the United States have helped install vehicle and personnel monitors, cameras, anti-tampering devices, and inventorying devices to improve security.[68]

Zheleznogorsk, or Krasnoyarsk-26, is also a plutonium city, also with a population over 100,000, and also in Western Siberia.[69] It is located on the Yenesei River, which flows north from the southern part of Lake Baikal to the Kara Sea, and is about 50 kilometers (30 miles) down river from much larger, civilian Krasnoyarsk. The Mining Chemical Combine continues to operate one of the three remaining plutonium production reactors. It is also the proposed site of a large reactor-fuel reprocessing plant, one that would take civilian fuel and recycle both uranium and plutonium for VVER-1000 power reactors.

Ozersk has been described by physicist Tom Cochran of the Natural Resources Defense Council as "probably the most polluted place on earth." The Mayak plutonium production facility there was built in 1948, with a nuclear fuel reprocessing facility added in 1977. Between 1948 and 1953, nuclear waste, including plutonium, was dumped directly into the Techa River, and later into the small Lake Karachai. Forty years after the waste was dumped, radiation monitors on the banks of the Techa indicate readings of 0.4 millirem per hour. A human living on the bank of the river would receive over 3.5 rem just from direct radiation from the wastes accumulated in the river. Dams, in questionable condition, block the flow of radionuclides down the Techa.[70] A notorious explosion at the Mayak facility in 1957 ejected 2 million curies of radiation from a tank containing reactor waste and hot, explosive salts. The material was thrown 1,000 meters into the air, resulting in the evacuation of 10,000 residents. The accident was kept secret until 1988.[71]

A serious question remains regarding Russian plans for large-scale plutonium reprocessing from spent reactor fuel. Controlling plutonium diversion would be exceedingly difficult in a large reprocessing facility—one handling 100 tons of spent fuel per year—because the margin of accounting error would be larger than the amount required to make a weapon.[72] Minatom nevertheless plans to build a plutonium reprocessing plant in Zheleznogorsk capable of processing over 100 tons of spent reactor fuel annually and recovering some 15 tons of plutonium. Financing would be secured by customers wanting recycled fuel, including utilities in South Korea, Switzerland, and Taiwan. Russian leaders cite a precedent for this approach in that British and French reprocessing facilities have been financed in part with fees paid by customers in Germany and Japan.

A related issue is the disposition of existing stocks of weapons-grade material. The United States launched a major effort to eliminate highly enriched supplies of uranium 90 percent fissile U-235–by diluting it to 4.4 percent U-235 by blending it with non-fissile U-238. The United States planned to finance the acquisition of this stockpile by selling the blend as reactor fuel for civilian light water reactors. An agreement was reached

for the United States to buy 500 tons of this bomb-grade material from
Russia at a value exceeding $10 billion.

The agreement has been frustrated by privatization—in the United
States. The U.S. government privatized its state-owned uranium fuel monopoly, the U.S. Enrichment Corporation (USEC), which had been given
responsibility to acquire Russian uranium for blending with depleted or
low-enriched uranium suitable for use in U.S. reactors. Having become a
for-profit firm, with responsibility to shareholders, the USEC was accused of viewing the Russian nuclear fuel surplus as a steel company
might see Russian imports, as a form of "dumping" threatening to drive
down prices and profits for the company's own product. The firm was
accused of reneging on a deal to purchase 18 tons of highly enriched uranium and instead offered to buy only 12 tons, a difference equal to several hundred bombs.[73] USEC maintains that the agreement was jeopar-
dized not by its concerns but by Russian misunderstanding of the terms
and schedule of payments, and that it has lived up to the letter and spirit
of the agreement. Even if the purchase agreement were implemented in
good faith, the complete purchase and disposition could require two
decades to complete.

Theft and diversion of nuclear materials will long remain the greatest
threat to both Russia and the United States. The key to avoiding such
theft is to safeguard the material at the source. It would be very difficult
to intercept nuclear material smuggled into the United States, or any
other country. Hundreds of thousands of tons of freight and millions of
people arrive in western countries each year. Thousands of tons of illegal
drugs each year make it through porous borders, and drugs in many
ways are easier to detect. Delivery of a terrorist weapon could be by air,
missile, or Ryder truck. Star Wars missile defense technology would be
useless against such a threat. But at the source, materials are difficult to
obtain. It is difficult to separate uranium and plutonium from production
reactors.

The U.S. government has long drawn a line between military and civilian nuclear technologies, believing that mixing the two would increase
the chances of proliferation. For one thing, allowing this practice would
undermine efforts to impose strict controls on non-nuclear nations. That
is, arguing that Iran should not be able to separate plutonium from spent
fuel even though the west did so would be too hypocritical. More important, allowing the separation and transport of plutonium would make far
more difficult the job of preventing weapons-usable material from being
diverted to clandestine uses. The creation of a plutonium economy has,
in the U.S. arms control community, been equivalent to proliferation of
nuclear weapons.

Counter-proliferation efforts have scored some quick successes and begun longer-term efforts. One early success airlifted 5 kilograms of plutonium from a research reactor in Tblilsi, Georgia, to Dounreay, Scotland. The research reactor had for five years performed no inventory of its weapons-usable material, and Georgia continued to be unstable following a fierce civil war, especially with conflict continuing in the territory of Abkhazia. A U.S. military plane evacuated the material to a nuclear facility at Dounreay for safekeeping.[74] In another success story, dubbed Project Sapphire, the United States bought and airlifted from Kazakhstan 600 kilograms of highly enriched uranium to the United States to be converted to reactor fuel.

Application of fundamental security measures known as "Materials Protection, Control, and Accounting" (MPC&A) is crucial to avert nuclear materials theft.[75] As the Murmansk naval facility nuclear incident illustrated, safeguards can be extremely lax and inadequate in some Russian facilities. Priority security upgrades include, for example, the installation of materials detectors for staff and vehicles in nuclear facilities. Methods and equipment for MPC&A implementation were made available and installed in several Russian installations with funding from the "Nunn-Lugar" act, which authorized funding for the U.S. Department of Energy to support regional efforts. The U.S. government also worked with Kazakh authorities to secure poorly guarded stockpiles of bomb-grade nuclear material.

The U.S. government has also recognized the need to help the Russian military nuclear sector find new livelihoods. Such activities fit within the overall effort to reduce nuclear proliferation. The Russian government in 1998 increased security concerns by failing to pay many military and nuclear personnel. Alexander Lebed, the former general and then governor of Krasnoyarsk Oblast, even offered to take over operation of a nuclear weapons facility with unpaid workers, an offer which, serious or not, raised the specter of a Russia fragmented into nuclear states.[76]

A U.S. "Nuclear Cities Initiative" was approved in 1998 to help Russian "plutonium cities" diversify their economies by training local leaders in economic development activities. The idea is to help connect the nuclear cities to business opportunities that would utilize their scientific and technical strengths, thus reducing the risk that workers would steal and sell nuclear material out of economic desperation. The program will provide training in marketing and business plan development and encourage the kind of activities that a "chamber of commerce" or community development organization would provide for isolated but talented communities in the United States. Some of the U.S. weapons and scientific communities have had to make similar transitions, although they had the advantage of being integrated into one of the world's largest markets with a substantial legal, financial, transport, and communications infrastructure. Trying to replicate their successes in a nation suffering from one of the greatest economic depressions in history is a formidable challenge.

A more controversial program would involve use of "mixed-oxide fuels" to dispose of Russian military plutonium by burning it in western power reactors.[77] The idea is that blending excess fissile material with commercial nuclear fuel would burn it up and thus eliminate the need to store it, essentially forever on human time scales. This approach would also raise revenues through fuel sales to help pay the cost of the effort. Alternatively, the fissile material must be guarded carefully or buried, perhaps fused in glass, and kept out of the hands of persons who would use it to make weapons. The outcome of this debate will hinge on U.S. willingness to use military materials in commercial applications—long a taboo—and on technical questions about the feasibility of long-term storage.

Although cooperation between the United States and Russia has been positive in the field of MPC&A, it has been less so with respect to Minatom's exports of nuclear facilities and equipment. Minatom's desire for international business has been an irritant in U.S.-Russian relations, particularly the agency's marketing and sales of reactors for export to Iran, China, and India.[78] The 1992 non-proliferation treaty signed by the Soviet Union, the United States, and much of the rest of the world prohibits the transfer of dual-use—useful for weapons as well as civilian purposes—nuclear technology to non-nuclear states. Minatom views the future of its complex increasingly as one of international commercial nuclear power, and has increasingly marketed its product around the world, legally, it argues. Minatom has exported as much as $2 billion per year in nuclear technology and fuel and hopes to double this rate early in the next century. It has reached agreement to sell China two light water reactors for use in a plant at Jianyungang in Jiangsu province for approximately $3.5 billion.[79]

However, U.S. foreign policy places high priority on non-proliferation of weapons-making capabilities. In particular, U.S. intelligence services warn of Iranian efforts to develop weapons of mass destruction, including nuclear weapons. Thus, Minatom's most contentious deal is to build an $800 million VVER facility at Bushehr, Iran, on the Arabian Sea, for which Iran provided a 10 percent down payment in the mid–1990s.[80] Need for the nuclear plant is difficult to understand on the basis of energy economics. Why, security specialists ask, would Iran acquire a $1 billion Russian nuclear power plant when it possesses immense quantities of natural gas, unless the Iranian regime wants to make an atomic

---

BOX 3.3   MAKING NUCLEAR WEAPONS MATERIAL

Making plutonium or highly-enriched uranium for weapons can be accomplished by physically separating fissile uranium from natural uranium, or breeding plutonium from uranium then separating it chemically. These processes require large-scale, industrial operations.

Uranium separation: Fissile atoms will fission after absorbing neutrons. Fertile atoms will transmute into fissile ones after absorbing neutrons. The element uranium has both fertile and fissile isotopes, including the isotope uranium 235, which will readily fission when bombarded with neutrons and uranium 238 which will not. The numbers represent the total number of neutrons plus protons in the nuclei of the two isotopes. The atomic number–the number of protons–in uranium is 92, and the number of neutrons for uranium 235 and 238 is 143 and 146, respectively. Uranium ore contains about 99.3 percent uranium 238, and 0.7 percent uranium 235. Enrichment of uranium 235 to 3 percent is necessary for power reactor fuel. Enrichment to 90 percent is necessary for military use, although a terrorist might make a destructive, if inefficient, weapon work using only half that concentration. Gaseous diffusion enrichment is very expensive, requiring that uranium hexafluoride be gasified and pressurized against a metal barrier that has molecule-sized holes. The barrier allows smaller molecules containing fissile uranium 235 to pass but not the larger ones containing uranium 238. An industrial scale plant might cost a billion dollars to build and require 2 GW of power to operate. Gaseous centrifuges, however, now can do the job for less money and much less power. This process also exploits the slight difference in mass by spinning the molecules because the heavier isotope tends to travel farther from the center of the centrifuge.

Breeding: Fertile atoms can be transmuted into fissile ones. In a reactor that produces neutrons, uranium 238 will absorb the neutrons and change to plutonium. This takes place in stages:

$$^{238}U_{92} + n \rightarrow {}^{239}U_{92} \rightarrow {}^{239}Np_{93} \rightarrow {}^{239}Pu_{94}$$

The first step has a half-life of about 24 minutes, the second step 2.4 days. Plutonium will decay to uranium 235 with a half-life of about 24,000 years. Seven half lives are required to complete transformation of 99 percent of the material.

Plutonium separation: Chemicals such as carbon tetrachloride can be used to separate plutonium from uranium. Military applications as well as reactors require a pure concentration of plutonium 239 because other isotopes of

(continues)

(continued)

plutonium will absorb neutrons and reduce the power of the chain reaction. Although every power reactor produces some plutonium, terrorists would have great difficulty separating it from spent reactor fuel because of the intense radiation in the fuel assemblies. A processing plant capable of separating enough plutonium for one or two bombs per year would cost a few tens of millions of dollars. India produced plutonium sufficient to build a nuclear weapons arsenal. American policy makers oppose Russian sale of power reactors to Iran to prevent that country developing nuclear weapons, and also oppose plutonium recycling generally because it would produce large quantities of plutonium that would be difficult to safeguard.

Crude Bomb-making: Making a crude bomb from weapons material would be relatively easy. It could probably be accomplished by skilled workers who had never made a nuclear weapon before. A device capable of devastating Manhattan would weigh a few tens of kilograms (around 100 pounds), and could be delivered in the back of a small truck. If a 1 kiloton device exploded in a city, it would produce an increase in pressure of roughly one-third atmosphere (5 pounds per square inch, or psi) over a radius of several square blocks. Increased pressure of 0.2–0.3 atmosphere (2–3 psi) could blow occupants out of office buildings, and many deaths would result from flying glass and debris, and from the collapse of buildings. The force of such a bomb would completely destroy light commercial buildings and homes, and would heavily damage office buildings. Worse, lethal radiation of 600 REM would be expected over a radius of 0.8 kilometer (0.5 mile). A badly made bomb from reactor-grade plutonium could "fizzle," go critical and explode before the reaction had time to produce a profound nuclear explosion. Nevertheless, it could still explode with the force of 1,000 tons of TNT. That force can be compared with the explosion that crippled the World Trade Center–one half ton of chemical explosive–or the Oklahoma City bombing, which represented less than two tons of fertilizer explosive. A terrorist's nuclear fizzle would have destroyed most of Oklahoma City, not just the Murrah Federal Building.

bomb? After all, Iran had tried to procure the necessary components to do so in the early 1990s. Although Iran already has a small research reactor capable of producing fissile material, it would take ten years of production to make enough plutonium for a single bomb. Ironically, this facility was provided by the U.S. government during the tenure of a different ruler.[81]

Minatom argues that international commercial nuclear power trade is legal, that safeguards are provided, and that the business helps to stabilize dangerous nuclear communities in Russia. Western response has alternated between the stick and the carrot, between pressuring the Russian government to suspend sales of dual use technologies to risky regimes and to be more responsible about non-proliferation, and supporting the commercial nuclear power enterprise in Russia and across Eastern Europe. Ironically, as Russia attempted to make a transition to market economics and as commercial nuclear power appears to be failing to compete in most markets, western assistance has helped shore up the commercial nuclear enterprise in Russia, and, by easing the financial burden of improving the safety of its own reactors at home, to indirectly and unintentionally foster the export of nuclear technology to questionable customers.

Conclusions

The clearest nuclear imperative in the transition economies is the safe, secure disposition of fissile uranium and plutonium. If the west helps with this task, it protects its own vital interests, especially as economic crisis continues to increase the risk that weapons-grade material could be diverted. East-west cooperation in nuclear materials protection and control has already achieved important successes by helping remove quantities of fissile material from unsafe locations. New programs to support the economic development of the plutonium cities may help improve nuclear materials security in the longer term.

Western influence over Russian nuclear technology exports has been minimal. Russia has been impervious to U.S. pressure to suspend reactor sales to Iran and India.[82] Western influence even in its best-funded area— power reactor safety—has been modest. Progress has been made in reducing fire risks and operator error through training and acculturation of technical workers, but efforts to close down unsafe reactors have been rebuffed. Because assistance programs have continued anyway, the west sends mixed signals. At the same time that the west claims that its main objective is to build a "safety culture," it quietly acquiesces in the continued operation of unsafe plants, thus signaling that this is "okay." No "safety culture" would allow operation of plants that significantly jeopardize the property, safety, and lives of regular citizens. A more balanced approach, one more likely to succeed, would change the balance of nuclear power, fossil, and demand-side assistance to aim for true reform in the energy sector. This change could mean increasing non-nuclear assistance, but more aid in the current political environment is unlikely, suggesting that a reorientation to achieve balance is needed.

Less urgent issues such as nuclear waste cleanup, associated with both military and commercial operations, may be beyond anyone's ability to do more than stabilize some of the more dangerous storage facilities. Many billion dollars have been spent on the Hanford Reservation in the United States for cleanup purposes, with controversial results. In the impoverished transition nations where so many needs go unmet, there may be little choice in the short term but to close and secure these areas.

The influence of foreign governments is always limited by the good will of local ones and the extent to which it can be curried with grants, loans, subsidies, and favors.[83] A not-entirely-abstract problem for western governments is whether use of public funds and guarantees to achieve private commercial goals, or even worthier ones related to safety and security, can do more harm than good by distorting the markets that everyone agrees the transition economies should develop. The danger is that western soft money comes to substitute for the subsidies previously provided under central planning, undermining privatization and market mechanisms in the process. Where western interests have been clearly perceived, there has been success in collaborating with the region's nuclear sector. United States administrations persuaded Ukraine and Kazakhstan to remove nuclear weapons from their soil, a contribution of incalculable value. Senators Richard Lugar and Sam Nunn persuaded Congress to provide substantial U.S. funds to help reduce the threat of plutonium diversion, leadership that provided perhaps the best U.S. security investment in the last quarter century. Where western interests have been less compelling, such as in the area of commercial nuclear deals, the results have been less clear.

For citizens of the transition economies, nuclear safety is a question of governance. Market reform is one answer because competitive markets can best choose power sources, provided that environmental externalities are accounted for. Environmental health and safety regulation is thus another answer, because the risks to the public are real, and every government has the responsibility to protect its citizens. Citizens of the region must build democratic institutions with the skill and authority to arrest irresponsible behavior while permitting entrepreneurship. Governments must allow markets to work without propping them up with protection, subsidies, and a license to take whatever risk they wish to accomplish their goal, be it the generation of electric power or jobs.

Notes

1. This text uses the Russian spelling because it was the name of the reactor when it was built and when the accident that made it famous occurred. The Ukrainian spelling is *Chornobyl*.

2. "Ukraine Says It Has Spent $14 Billion on Chornobyl Cleanup," *RFL/RL Newsline* 1., no. 18, pt. II (24 April 1997); http://www.rferl.org/newsline/1997/04/240497.html.

3. Robert E. Ebel, *Chernobyl and Its Aftermath: A Chronology of Events* (Washington, D.C.: The Center for Strategic & International Studies, 1994), 2.

4. Robert Ebel at the Center for Strategic and International Studies estimates that 237 persons received more than 100 REM–that is, became acutely ill with radiation syndrome—and that 28 persons received lethal doses. Of those who became sick, 209 were cured, and 13 were incapacitated. Today at Chernobyl, if one lived 24 hours a day 365 days a year in Reactor Building number 1, and the rate remained constant, one would get approximately 3 REM per year, below the maximum occupational exposure limit of 5 REM per year.

5. Michael Balter, "Children Become the First Victims of Fallout, Special News Report," *Science* 272 (19 April 1996): 357; Nigel Williams, "Leukemia Studies Continue to Draw a Blank," *Science* 272 (19 April 1996): 358.

6. Lydia Popova, Social-Ecological Union, personal communication, Moscow, October 1998.

7. About 15.4 pounds.

8. The same would be true of a nuclear weapon, except that bomb designers found ways of reflecting neutrons back into the fissile material and forcing a "super-fast prompt critical" chain reaction, a nuclear explosion. Such an explosion is virtually impossible in a reactor because it lacks the exacting geometry and materials necessary to produce the explosion. The Chernobyl accident produced an explosion that was orders of magnitude less powerful than even a small weapon. The reactor was destroyed with a steam explosion and continued to release radio nuclides for days mainly as a consequence of a fire–a chemical reaction–in the graphite moderator. Note that graphite is flammable.

9. No U.S. plutonium production reactor was operating at this writing.

10. Roughly 250 pounds.

11. Recall from basic physics, however, that uranium 238 is "fertile," meaning that each nucleus can absorb a neutron and transmute to fissile plutonium 239.

12. For a useful perspective by a nongovernmental organization, see Tim Jenkins, *Dangerous Liaisons: Western Involvement in the Nuclear Power Industry of Central and Eastern Europe* (Washington, D.C.: Friends of the Earth, January 1993).

13. This total includes reactors actually operating in 1997: fourteen RBMKs; ten VVER 440-230s; one BN-350 (fast neutron or breeder reactor); one BN-600; and four LWGR-12, light water graphite reactors, similar to RBMKs except much smaller. See Office of Nuclear Energy, Science and Technology, *Improving the Safety of Soviet-Designed Nuclear Power Plants,* U.S. Department of Energy (Washington, D.C., 1 November 1996).

14. The so-called LD 50/30, the "lethal dose for 50 percent of any given population within a period of 30 days," is 500 REM. Radiation sickness (nausea, loss of hair, changes in the blood) occurs at about 100 REM, and the maximum dose for professional nuclear workers is 5 REM per year. Natural background radiation is around 0.2 REM (200 millirem), and a full-mouth set of dental X-rays would expose one to about 50 millirem (probably less with modern film and well-trained specialists).

15. Vladimir Udovichenko, personal communication, Slavutych, Ukraine, February 1998.

16. Only one unit was operating. In addition to the unit destroyed in 1986, another was previously damaged by fire. Closure was scheduled for late 2000.

17. Science Policy Research Unit, University of Sussex, *Economic Assessment of the Khmelnitsky 2 and Rovno 4 Nuclear Reactors in Ukraine: Main Report,* by John Surrey et al., draft report to the European Bank for Reconstruction and Development, January 1997. See also Simon Holberton, Leyla Boulton, and Matthew Kaminski, "Chernobyl Closure Plan Hits Snag," *Financial Times,* 8/9 February 1997, p. 1; Leyla Boulton and Simon Holberton, "West's Policies on Eastern Nuclear Plants 'Misguided'," *Financial Times,* 10 February 1997, p. 2; and Sander Thoenes, "Risk of Delay in Chernobyl Shut-down," *Financial Times,* 4 July 1997, p. 3.

18. Holberton, Boulton, and Kaminski, "Chernobyl Closure Plan Hits Snag," 1.

19. Thoenes, "Risk in Delay in Chernobyl Shut-Down," 3; Tony Wesolowsky, "EBRD Plans to Fund Controversial Reactors in Ukraine," *RFE/RL Newsline* 3, no. 33, pt. II (17 February 1999); http:/www.rferl.org/nca/featrues/1999/01/f.ru.990129145538.html; Charles Clover and Ralph Atkins, "Ukraine to Ponder Non-Nuclear Option," *Financial Times,* 10/11 July 1999, p. 11.

20. "G-7 Offers $300 Million to Rebuild Chornobyl Sarcophagus," *RFE/RL Newsline* 50, pt. II (23 June 1997); http://www.rferl.org/newsline/1997/06/230697.html.

21. Oleg Varfolomeyev, "Ukraine Gets Nuclear Fuel," *OMRI Daily Digest* no. 60, pt. II (26 March 1997); http://www.rferl.org/newsline/1997/03/3-CEE/cee-260397.html.

22. Mark Tomlinson et al., "Ukraine Nuclear Power Sector Fact-Finding Mission on Nuclear Safety, 5–8 April 1994" (World Bank and European Bank for Reconstruction and Development, 15 April 1994, mimeographed).

23. Of Ukraine's fourteen operable reactors, only four were considered highly unsafe: two VVER 440-230 reactors and two RBMK units at Chernobyl.

24. "G-7 Offers $300 Million to Rebuild Chornobyl Sarcophagus."

25. For nonsecurity reasons, The Office of International Nuclear Safety, which managed the Soviet nuclear safety account, was moved to the Office of Nonproliferation and National Security. See "DOE News," Department of Energy press release, 8 May 1998; and Lira Behrens, "Peña Orders Lash Out as Director of NE," *Inside Energy,* 11 May 1998, pp. 1 and 11–12.

26. Lithuania, of course, uses little power in comparison to larger countries.

27. This amount of energy is roughly equal to that provided by a moderate-sized oil pipeline carrying about 750,000 barrels per day.

28. Strikingly, the largest populations placed at risk by these RBMKs are not in Ukraine but at St. Petersburg in Russia and Ignalina in Lithuania. The Leningrad Nuclear Power Station is situated about 55 kilometers west of St. Petersburg's urban population of 4 million people. The Chernobyl exclusion zone, an area where now no one is permitted to live, has a radius of 30 km. A panel of EBRD safety experts recommended against the continued use of the two Ignalina reactors.

29. This problem extends to scientific and military facilities. Safety reviews were ordered in July 1997 by the Ministry of Atomic Energy after Alexander Zakharov died three days after exposure to 1,000 roentgens at Arzamas-16, a Nizhny Novgorod nuclear experimental facility. Associated Press, Moscow, as reported in "Radiation Checks Ordered," *Financial Times,* 2 July 1997, p. 2.

30. Matej Vipotnik and Robert Anderson, "Lithuania To Close First of Two Reactors by 2005," *Financial Times*, 9 September 1999, p. 2.

31. Michael Specter, "Occupation of a Nuclear Plant Signals Russian Labor's Anger," *New York Times*, 7 December 1996, pp. 1 and 6. The Minatom spokesman quoted was Georgi Kaurov.

32. "Yeltsin Orders Minister to Pay Nuclear Plant's Workers," *RFE/RL Newsline* 1, no. 76, pt. I (18 July 1997); http://www.rferl.org/newsline/1997/07/180797.html.

33. Tomlinson et al., "Ukraine Nuclear Power Sector Fact-Finding Mission on Nuclear Safety."

34. The author managed this program for the U.S. Department of Energy and the U.S. Agency for International Development.

35. Manfred Banaschik, European Bank for Reconstruction and Development, "The Activities of the Nuclear Saftey Account of the EBRD" (statement before the European Parliament, Public Hearing on East-West Cooperation—Energy and Nuclear Security, Brussels, 25 June 1996, mimeographed), 2-3.

36. In Bulgaria.

37. A panel of EBRD experts recommended that Ignalina not be restarted after a planned maintenance outage in 1997. The panel conducted a review of both design and operational issues, and harshly criticized the Lithuanian regulatory agency VATESI and the plant operators, stating that they did not "fully understand and accept their respective roles and responsibilities to resolve safety issues. . . . If their roles and responsibilities are not clearly defined, understood, and aggressively implemented, this would be a significant issue requiring immediate shutdown." "Restart of Lithuanian Reactors Opposed by European Bank Panel," *Post-Soviet Nuclear & Defense Monitor* 4, no. 7 (7 April 1997): 4.

38. Boulton and Holberton, "West's Policies on Eastern Nuclear Plants 'Misguided'," 2.

39. Paxus Calta, Nuclear Information and Resource Service (NIRS), "Nuclear Politics," letter to the editor, Washington, D.C., 2 May 1998.

40. Interestingly, some of the sharpest criticism has come from groups such as the Natural Resources Defense Council, from which several positions of leadership in the Clinton Administration had been filled. That is, their criticism was not blunted because those being criticized were their former colleagues.

41. Names of the closed cities housing the populations working at Chelyabinsk-65, Tomsk-7, and Krasnoyarsk-26, respectively.

42. Natural Resources Defense Council, "Results of the Moscow Nuclear Safety Summit" (Washington, D.C., April 1996, mimeographed).

43. Robert Orttung, "Krasnoyarsk Considers Nuclear Referendum," *OMRI Daily Digest* I no. 51 (13 March 1997); http://www.rferl.org/newsline/1997/03/1-rus/rus-130397.html.

44. Although Minatom has marketed heavily to export reactors to China, India, and Iran, even if completed these sales cannot make up for the losses of the Russian nuclear power enterprise.

45. See, for example, the "Joint Alternative Electric Energy Strategies Study." This $8 million report was funded by the United States under the umbrella of the

Gore-Chernomyrdin Commission to identify ways of satisfying Russian electric power demand in the future in a safe and clean manner.

46. A State Department spokesperson, Carol Kessler, in 1994 proposed to limit liability for international investors in Russia. See Carol Kessler, "The Role of Foreign Assistance in Improving Nuclear Safety," *CIS Environmental Watch* 7 (Fall 1994/Winter 1995): 60–69 (proceedings of the Symposium on Nuclear Safety in the Former Soviet Union, Monterey, Calif., 12–13 April 1994).

47. A Three-Mile Island reactor unit recently sold was one not damaged in the famous accident.

48. Westinghouse purchased a 33 percent share in a joint venture company named I&C Energo. The Czech partners are Ceske Energeticke Zavody (CEZ), which owns 34 percent, and CME of Trebic (Moravia), a 33 percent owner. I&C Energo will complete safety upgrades of Temelin. "Czech Republic: Joint Venture Formed," *Nucleonics Week* 34, no. 31 (5 August 1993): 15.

49. Ed Lane, "Westinghouse Sees Potential in East Europe," *The Energy Daily* 21, no. 130 (9 July 1993): 3–4.

50. "Senate Committee Request for Temelin Information Not Disapproval," *CTK National News Wire*, 29 July 1993; "American Bank Says Asking for Temelin Details Is Not Exceptional," *CTK National News Wire*, 29 July 1993; "Survey of Czech Press," *CTK National News Wire*, 3 August 1993.

51. Citibank Agrees to $400 Million Loan for CEZ," *Reuters European Business Report*, 30 July 1993.

52. "Division in CSFR Electricity Sector Set to Intensify," *East European Energy Report*, 16 September 1992.

53. Personal and confidential communication, Prague, Czech Republic, 5 December 1995.

54. The plant owner and operator would be Slovenske Elektrame, a state-owned monopoly.

55. Reuters, "Slovakia Warned Over N-Plant," *Financial Times*, 20 May 1998, p. 3.

56. See generally Jae Edmonds and John M. Reilly, *Global Energy: Assessing the Future* (New York: Oxford University Press, 1985). See also J. A. Edmonds, *A Guide to Price Elasticities of Demand for Energy Studies and Methodologies,* research memorandum ORAU/IEA–78–15(R), Institute for Energy Analysis (Oak Ridge, Tenn., 1978); D. R. Bohi, *Analyzing Demand Behavior: A Study of Energy Elasticities* (Baltimore, Md.: Johns Hopkins University Press for Resources for the Future, 1981); W. D. Nordhaus, *International Studies of the Demand for Energy* (Amsterdam: North-Holland, 1978) (selected papers presented at a conference held at the International Institute of Applied Systems Analysis); Energy Modeling Forum, *Aggregate Elasticity of Energy Demand*, Vol. 1 (Stanford, Calif., 1980); *Report of the Working Group on Energy Elasticities*, Energy Paper 17 (London: Her Majesty's Stationery Office, 1977).

57. Nuclear Material Security Task Force, *United States/Former Soviet Union Program of Cooperation on Nuclear Material Protection, Control, and Accounting*, U.S. Department of Energy, Office of Arms Control and Nonproliferation (December 1996).

58. Russian experts and politicians express doubt about certain episodes, particularly incidents in Europe, citing evidence that supposedly shows western intelligence agents helped stage the events.

59. This discussion summarizes an excellent presentation in Graham T. Allison, Owen R. Coté, Jr., Richard A. Falkenrath, and Steven E. Miller, *Avoiding Nuclear Anarchy* (Cambridge, Mass.: MIT Press, 1996).

60. See generally Thomas Cochran, Robert Norris and Oleg Bukharin, *Making the Russian Bomb* (Boulder, Colo.: Westview Press, 1995).

61. Allison et al., *Avoiding Nuclear Anarchy*. The reader should be aware that Russian observers, including some experts, regard this and other episodes involving western security forces as suspicious and politically motivated.

62. U.S. Delegation to Tomsk and Krasnoyarsk-26, *Replacement Power Fact Finding Mission for Plutonium Producing Reactors*, trip report for 7–16 June 1994, U.S. Department of Energy, Office of Nuclear Energy (Washington, D.C., September 1993).

63. Excellent source books on these issues are David Albright, Frans Berkhout, and William Walker, *Plutonium and Highly Enriched Uranium 1996: World Inventories, Capabilities, and Policies* (Oxford: Oxford University Press, 1997); and Don J. Bradley, *Behind the Nuclear Curtain: Radioactive Waste Management in the Former Soviet Union* (Columbus, Ohio: Battelle Press, 1997).

64. Samuel Glasstone and Philip J. Dolan, *The Effects of Nuclear Weapons* (Washington, D.C.: U.S. Department of Defense and the Energy Research and Development Administration, U.S. Government Printing Office, 1977).

65. The U.S. Congress approved funding to aid Russian "Plutonium Cities" in 1998, and U.S. Vice President Al Gore signed an agreement with Russian Prime Minister Kiryenko to implement it.

66. Note that this production is done by agreement with the U.S. government to reduce the threat of plutonium diversion.

67. Lydia Popova, Social-Ecological Union, personal communication, Moscow, October 1998. See also Eduard Kayukov, Wendy Wallace, M. Carolina Escalona, and Kristen Wesenberg, "Tomsk–7: Anatomy of an Accident," *CIS Environmental Watch* 4 (Summer 1993): 57–65.

68. Nuclear Material Security Task Force, *United States/Former Soviet Union Program of Cooperation on Nuclear Material Protection, Control, and Accounting*, U.S. Department of Energy, Office of Arms Control and Nonproliferation, L-L–43–46 (December 1996).

69. Not to be confused with the uranium enrichment facilities at Zelenogorsk, or Krasnoyarsk-45.

70. Peterson, *Troubled Lands*, 148–149; Lydia V. Popova et al., *Plutonium in Russia* (Moscow: Social-Ecological Union), 1994), 40, in Russian.

71. Don J. Bradley, *Behind the Nuclear Curtain: Radioactive Waste Management in the Former Soviet Union* (Columbus, Ohio: Battelle Press, 1997).

72. U.S. Congress, Office of Technology Assessment, *Nuclear Safeguards and the International Atomic Energy Agency*, OTA-ISS–615 (Washington, D.C.: Government Printing Office, April 1995), 3.

73. Peter Passell, "Profit Goals May Hobble U.S. Atomic Fuel Agency," *International Herald Tribune*, 29 August 1996, p. 1.

74. Michael R. Gordon, "U.S. and Britain Relocate a Cache of Nuclear Fuel," *New York Times*, 21 April 1998, p. 1. Strong negative public reaction followed in Scotland, perhaps making the replication of this approach unlikely. See Chrystia Freeland, "Scientists Warn of More Nuclear Perils," *Financial Times*, 23 April 1998, p. 2.

75. Office of Nonproliferation and National Security, *MPC&A Program Strategic Plan*, 2–3, U.S. Department of Energy (January 1998).

76. Chrystia Freeland, "Lebed Raises Specre of Nuclear-Armed Provinces," *Financial Times*, 25/26 July 1998, p. 1.

77. U.S. Senate, S. 2138, 105th Cong., 2d Sess., 5 June 1998, S. Rept. 105–2060. See also Office of the Vice President,*U.S.-Russian Agreement on Management of Used Plutonium: Agreement Announced at Gore-Kiriyenko Working Meeting 7/23–24*, White House (23–24 July 1998).

78. Minatom would point out that U.S. firms also market heavily abroad, including in the transition region, and that exporting to Iran and China is not prohibited under the non-proliferation treaty. Regarding the first point, see Ed Lane, "Westinghouse Sees Potential in East Europe," 3–4. See also "Russia, India Sign Preliminary Contract on Nuclear Plant," *RFE/RL Newsline* 2, no. 138, pt. I (21 July 1998); http://www.rferl.irg/newsline/1998/07/210798.html.

79. "Russia and China to Sign Nuclear Cooperation Document," Itar-Tass News Service, 24 June 1997; Robert Orttung, "Moscow Holds Firm on Reactor Sale to India," *OMRI Daily Digest* no. 29, pt I (11 February 1997); http://www.friends-partners.org/friends/news/omri/1997/02/970211I.html; Jeffrey Logan, Director for China Programs, Battelle, Pacific Northwest National Laboratory, personal communication, Washington, D.C., October 1998.

80. *Interfax*, 21 March 1997.

81. David A. Schwarzbach, "Iran's Nuclear Puzzle," *Scientific American* 276 (June 1997): 62–65.

82. "Moscow Holds Firm on Reactor Sale to India," *OMRI Daily Digest* no. 29, pt. I (11 February 1997); http://www.rferl.org/newsline/1997/02/1-RUS/rus-110297.html; Reuters (wire service), "Russia to Push Forward with Iran Nuclear Reactors," 11 January 1999.

83. "Replacement Power for Pu-Reactors Almost Certainly Nuclear Based," *Post-Soviet Nuclear & Defense Monitor* 3, no. 3 (19 December 1995): 2–4.

4

Petroleum Geoeconomics

Geopolitics is an extravagant and therefore appropriate word to describe the struggle for oil and gas in Siberia and the Caspian Sea area and for the right to deliver fuel to the world.[1] Much is at stake: money, reform, power, security, development, environment. Petroleum exports are key to Russian economic recovery and transition to a market-based democracy, a vital goal for both east and west. Energy exports could help lift Central Asia and perhaps the Caucasus out of poverty. They could also diversify the west's oil options away from the Middle East. Stable, independent Central Asian nations could help anchor the region against extremism and control rogue states' efforts to develop weapons of mass destruction. The Caspian and Siberia remain great prizes for energy corporations, although the rest of the region has little oil and gas. Governments and companies using each other to accomplish their goals in the Caspian evoke the days of the East India Company, when it was "difficult to sort out whether the corporations are driving foreign policy, or the policy has marshaled the commercial interests behind it."[2]

Competing goals inevitably conflict. The west once ranked regional oil development as a priority for Russian economic recovery because oil and gas exports generate badly needed wealth for transition to democracy and markets. Indeed, low oil prices helped bankrupt Russia's economic stabilization program. However, the Russian backlash against capitalism has made the nation even less friendly to foreign investors at a time when U.S. commercial interests had already grown frustrated with Russia's opaque commercial law and confiscatory taxes. This frustration, coupled with recent oil and gas discoveries in the Caspian region, has shifted western attention to Central Asia. The U.S. and European governments have shifted their attention as well, seeking not only commercial advantage[3] but regional counterweights to Iran and Russia and energy independence from the Persian Gulf. Two wars in the nineties in the North Caucasus brought this division into sharp, bitter contrast.

Several policy contradictions jeopardize energy development in Russia and the Caspian. Promoting oil development with credits and implied military security increases downward oil price pressures in the short term, perhaps undermining the development schemes. Encouraging energy development is synonymous with encouraging consumption, an outcome inconsistent with western governments' commitments under the Framework Convention on Climate Change. Promoting energy development also conflicts with environmental conservation in the Siberian wilderness, on fragile tundra, and in pristine Pacific coastal areas, natural resources comparable to Alaska's North Slope, although some analysts hope that Western practices will improve on the careless approach taken by Soviet managers (see Figure 4.1). Promoting oil development may not even provide the social benefit sought for Central Asia; certainly it has not unambiguously improved the human condition in Angola, Iraq, Iran, Kuwait, Libya, Nigeria, or Saudi Arabia.[4]

Perhaps the most serious contradiction, however, comes in a policy of promoting Russian reform and economic transition on the one hand and aggressively competing with Russian commercial interests on the other. One oil industry analyst wrote succinctly that, "Resolving these two big problems [oil ownership and transport] will involve breaking Russia's domination of the Caspian."[5] Specialists in the United States maintain that this approach does not characterize their policy, that their goal is both Russian recovery and Central Asian independence.[6] Complicating these goals, which with regional cooperation might actually be compatible, is the fact that containing Iran ranks as a top U.S. priority, and frequent Russian-Iranian collaboration on Caspian issues creates for U.S. policy makers an awkward situation.

Although Caspian oil dominates headlines, Russian natural gas constitutes an energy resource of much greater physical, environmental, and even economic importance. Russia produces one-quarter of the world's natural gas. In energy equivalents, proven Russian gas reserves exceed proven oil reserves of Russia plus Central Asia by 500 percent.[7] The gas is controlled by one nation—indeed, one company—a cause of anxiety for democracy, security, and economics. The technical and environmental advantages of gas make it the fuel of choice for many European businesses and consumers. Leaders juggling issues of climate change, economic reform, and international security will want to look beyond near-term commercial interests and consider a wider set of energy choices.

Figure 4.1   Soviet-Era Oil Production, Baku, Azerbaijan, 1998.
Photo by William Chandler.

Shifting Caspian Fortunes

Oil was discovered bubbling to the surface on the Caspian at Baku a thousand years before the internal combustion engine was invented. Russia acquired this part of the Caucasus and access to the Black Sea from the Ottoman Empire under Nicholas I in the early nineteenth century.[8] Commercial production goes back over a century, when Baku became the second site in the world to produce commercial petroleum (after Ploiesti, Romania).[9] The Nobel brothers helped develop a thriving capitalist business there, only to have it expropriated after the October Revolution. The Caspian nevertheless produced most of Russia's oil during World War II. A decline in production saw a shift in the petroleum center of the region, first to the Urals, then to Western Siberia. Oil was only discovered in Western Siberia in 1960 and was not significantly produced for another decade. But lying under the permafrost were "supergiant" fields, including famous Somotlor. A supergiant is one containing more than 3.5 billion recoverable barrels; Somotlor contained an estimated 14 billion recoverable barrels. Now that Somotlor is exhausted, the

industry has looked again at the Caspian and has found resources totaling 178 billion barrels of oil.[10]

Energy analysts put such numbers into perspective by comparing
them to tangible things. For example, the United States in the late 1990s
consumed just over 6 billion barrels of oil per year (17.4 million barrels
per day, or 38 exajoules). Somotlor thus contained the equivalent of a little over two years of U.S. oil consumption. The United States uses tens of
thousands of wells to supply its oil on an annual basis, putting Somotlor's size into perspective. In its heyday in the mid-1980s, Somotlor could
have satisfied the oil needs of the entire Soviet Union for four and one-
half years.

Saudi Arabia, the United States, and Russia rank as the top three world
oil producers, in that order. But Russian oil production has in the last
decade declined by an amount greater than is produced by any country,
other than the top three. Production within the boundaries of the Russian
Republic fell from 11.4 million barrels of oil per day in 1988 to 6.2 million
barrels per day in 1997. The world's fourth largest producer, Iran, produces just 3.7 million barrels per day. Kuwait, when Iraq invaded, produced under 1.5 million barrels per day.[11] Russia achieved a small increase in production in 1997, the first in over a decade.

In the 1980s, the Soviet Union against great odds became the world's
largest oil producer. Soviet oil lay under the world's largest wetland,
frozen eight months per year and flooded most of the rest. Western Siberia dominated regional oil production for three decades. Even now, it accounts for 70 percent of regional output. All of Central Asia currently
produces only about 1.2 million barrels per day. Western Siberia is
drained by the Ob and the Yenisey Rivers, which flow north into the Arctic Ocean. The upper, more southerly portions of these rivers rise in the
Altai Mountains, which feed the rivers with snow melt and heavy rain.
The middle and lower, or northerly, sections of the rivers remain frozen
longer into spring, and ice dams cause spring floods along a broad flood
plain. The Vasunganye swamp that lies south of the Middle Ob and
Trans-Siberian Railroad alone covers more territory than New Jersey,
New York, and Pennsylvania combined.[12]

Communist planners financed their industry without private capital,
designed new drill technology, built cities to drill fields, and lay pipelines
on frozen tundra where everything—every meal, every cement block,
and every worker—had to be brought from somewhere else.[13]

Russia's oil decline has produced remarkably little impact on the
world's oil market. American officials have long worried about the oil
shock and security woes that would hit the world market if Saudi Ara-
bian production were suddenly lost. Russia's lost production over the
last decade is comparable with Saudi production, but in fact it has been

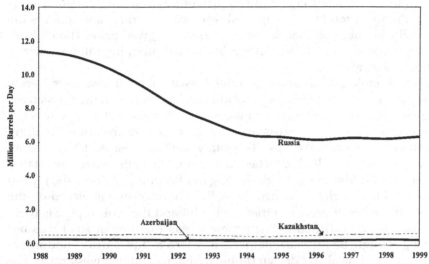

Figure 4.2   Oil Production, Largest Regional Producers, 1988–1999.
SOURCE: BPAmoco, *Statistical Review of World Energy, 2000,*
http://www.bpamoco.com/worldenergy.

offset by increasing output from the Middle East, especially from Saudi
Arabia. Central Asian production, meanwhile, is little changed from a
decade ago and still matters little to the world market (see Figure 4.2).

Oil Futures

Western Siberia's mammoth oil and gas fields accounted for the success
of the Russian oil industry. These fields ranked among the all-time greats
of oil and gas production. Somotlor field alone once produced 3 million
barrels per day, roughly equal to Mexico's current production. Somotlor,
now controlled by Nizhnevartovskneftegaz, has seen its total production
plummet. This trend, however, was predicted a decade ago by both Soviet and U.S. observers. Oil fields go through a production life cycle, and
it was anticipated that smaller fields would have to replace Somotlor.

Somotlor is a symbol of "socialist greed." Pumping rates were set at
unsustainably high levels. If oil production is not carefully regulated,
pressure in a field will decline and less oil will be recovered. Western
markets developed regulatory agencies to control over-exploitation, the
first and most significant being the Texas Railway Commission. That
agency prevented a tragedy of the commons by assuring that independent drillers pumping from the same field did not obtain an advantage

by pumping rapidly to get more oil than others pumping the same reservoir. Pumping too fast damages oil reservoir structures and reduces the overall amount of oil that can be recovered at a given price. The Railway Commission's job included setting individual pumping rates at socially beneficial levels.

The technological difficulties of the Russian oil enterprise were merely symptoms of the confused organization and incentives in the Russian energy sector,[14] but they were and are severe. Russian technology was not good enough to allow Russian oil drillers to use the rotary drilling method dominant in the west. The rotary method consists of high-quality steel pipe with a drill bit attached at the end. The entire assembly rotates, boring the bit into the ground. Drilling mud is pumped down the pipe to cool the bit and clean the drill hole. Enormous torque is placed on the pipe and, when it breaks, retrieving the bit and the broken pipe is time-consuming and difficult. The problem was that Soviet steel was not strong enough. So Russian engineers invented a "turbo-drill," finding solutions to technical and design problems that had eluded westerners. The turbo-drill had the advantage of allowing horizontal drilling, which saves time and money and reduces the impact of drilling on land because less surface area is disturbed by the rigs, which can drill laterally from the site. This alternative technology, however, had its own problems, such as the rapid wearing of the drill bit, which required pulling the entire length of the drill line and replacing the bit. Rotary drills must also be pulled—along with all the piping on top—but much less frequently. The problem is not that the Soviet technology will not work, but that it works more slowly and therefore costs more and reduces the amount of exploration that can be done with available teams, which in turn makes it hard to keep up with the pace of discovery necessary to replace the supergiant fields with more numerous and dispersed smaller fields. Doing so may simply not have been economical in the first place, particularly because the high rate of production was never really economic, being artificially stimulated by the command economy. Rationalizing the Russian oil industry may naturally—just as with metallurgy and other large industries—mean downsizing.

The infrastructure to develop and support oil exploration and production was an enormous burden both economically and environmentally. In remote areas of Siberia, not only was the work itself difficult, but problems were compounded by the need to provide housing and support for the workers. Workers typically are brought in shifts and live in dormitories.[15] These costs are a reason why some analysts estimated that renewing production at Somotlor would cost $6 billion.

Oil production is categorized in three stages: primary, secondary, and tertiary recovery. These stages refer to different means of getting the oil out of the pores in and spaces between rocks and bringing it to the surface. Sometimes the oil reservoir is under pressure from water below the oil, gas mixed in the oil, or gas trapped between the oil and rock between it and the surface. The pressure stems from the fact that in forming, oil and gas have usually risen toward the surface and sometimes displaced water from rock but get trapped and are unable to reach the surface. The aquifer below and gas below, within, and above can maintain pressure. Drilling a well into the underground reservoir will release that pressure under the hole, and gas and water pressure will push the oil in the reservoir toward and up the well. When that pressure is no longer adequate to maintain oil flow at the desired rate, workers can move to secondary recovery. Primary recovery will remove 10–50 percent of the oil in the reservoir.[16]

Secondary recovery usually involves pumping water and sometimes re-injecting gas into the reservoir to maintain pressure and to force the oil to flow to the surface. A common image of the oil industry everywhere is the rocking pump arm, which forces a kind of plunger—a sucker rod—down the well to pump oil. Sucker rods have to be used in secondary recovery when the pressure on the field is still not sufficient to bring oil to the surface even with water flooding. A more-efficient, higher-volume technology is the submersible pump, which is lowered down the well. These pumps provide greater volume and higher efficiency, but their quality in Soviet industry was not high.

The Soviet oil industry has been faulted for improper recovery techniques in using a process called water flooding. Almost all Western Siberian oil was produced with this technique. The Soviet industry argued that it was necessary to increase the ultimate recovery of oil and that it reduced the numbers of wells that had to be drilled and therefore reduced capital costs and time requirements. It would incidentally have reduced land impacts, if fewer wells were drilled. The water used in secondary recovery substitutes for the natural pressure of the reservoir when the pressure is no longer sufficient to push oil up the well to the surface. In Soviet practice, water was injected earlier than it would have been in the west. Western oil experts say that excessive water pumping damages the geological structure of reservoirs by breaking through impermeable barriers and overtaking the oil itself. Indeed, the share of water pumped with oil in Soviet fields increased from about 40 percent in 1965 by 10 percent every decade. West Siberian oil wells in the 1980s pumped liquid that was three-quarters water.[17]

Whether Russia's oil sector can recover or even sustain current levels of production is a question that requires distinguishing between reserves and resources. Resources is a theoretical category based on estimates from knowledge about geology. Reserves have been better defined, or identified, usually by significant exploratory drilling. The U.S. Geological Survey (USGS) defines reserves in an economic sense. Reserves include the amount of oil probably recoverable at current oil prices. The USGS further categorizes reserves as "proved, probable, or possible." Although the USGS approach provides a conservative estimate, it is consistent and understandable. Russian reserve estimates are more complicated and not so transparent. They are categorized as "explored reserves" plus "estimated" plus "prospective" plus "possible."[18] None of these categories has an explicit criterion for economic feasibility.[19]

The International Energy Agency (IEA), like most western oil specialists, discounts oil reserve estimates made using the Soviet system. The IEA estimates that Russia's resources amount to some 10 billion tons, or 70 billion barrels, the world's eighth largest national total.[20] The IEA believes the true number to be 50 billion barrels, although it acknowledges that the estimate is conservative and that the volume could be as great as 100 billion barrels. The IEA places 85 percent of former Soviet oil reserves in Russia, with two-thirds in Western Siberia alone. About one-fifth is in the Volga-Urals region, with Timan-Pechora near the Barents Sea at 10 percent and Lena-Tenguska at 3 percent. These latter areas are very sensitive environmentally. Western Siberia, being largely a wetland, may be just as sensitive. One-third of Russia's reserves are located in difficult areas—the Arctic and offshore—where environmental constraints could be high.

Outside of Russia and the Caspian area, countries of the region have little or no oil. Only Ukraine and Romania still produce significant quantities. No country of the region outside of Russia produced more than 0.5 million barrels of oil per day in 1999. Romania produced only 140,000 barrels per day, the energy equivalent of five large power plants. (For reference, total U.S. oil consumption in 1999 was over 17 million barrels per day.[21]) Ukraine's 1995 oil production amounted to only about 75,000 barrels per day.[22] Total Ukrainian reserves are minimal, on the order of 1.7 billion barrels. Moreover, the oil lies some 4–7 kilometers (2.4–4.2 miles) deep, in small deposits and complicated geology, and is viscous and high in sulfur.[23] Ukraine also produces about 20 billion cubic meters of gas (0.77 exajoules) per year.[24]

The best sense that analysts have been able to make of the estimates is to conclude that Russia's reserves outweigh those of the Caspian region by five or six or more to one, but that they pale in comparison to the Middle East, where Saudi Arabia alone probably holds four times as much recoverable oil as the entire former Soviet Union.[25] However, the Russian reserves are sufficient to drive an industry capable of producing 35–90 billion barrels of oil (210–545 exajoules), enough to provide the region with oil at mid-1980s consumption levels for ten to thirty years. These es-

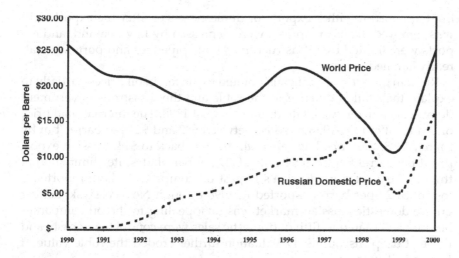

Figure 4.3   Russian Versus World Crude Oil Prices, 1990–2000.

SOURCE: E. M. Khartukov, "Low Oil Prices, Economic Woes Threaten Russian Oil Exports," *Oil & Gas Journal* (8 June 1998): 26; and Mark A. Gyetvay, "Restructuring, Consolidation Top Solutions for Russia's Major Oil Companies' Woes," *Oil & Gas Journal* (13 March 2000): 23.

timates are based, more or less, on current oil prices, and could grow significantly if prices rise significantly. The numbers are in any case large enough to make the region a target of opportunity for years to come, regardless of price.

Russian petroleum economist Eugene Khartukov notes that while the world oil price has been on a rollercoaster, Russian domestic oil prices have been flat (see Figure 4.3). He attributes this in part to the way price liberalization was handled for crude oil and for oil products. Crude oil prices were controlled until mid–1993, but product prices had been freed, had soared, and had started driving down demand for oil products, and in turn for crude oil. Crude oil prices, adjusted for inflation, were only one-quarter world price levels. Price-induced demand reduction meant that oil consumption in Russia was far below refinery capacity. Markets cleared with low relative prices. The price of crude oil sold for domestic use in Russia has not risen above half the world market level since the fall of the Soviet Union.[26]

Russia's price liberalization also had consequences that surprised the international community. In particular, the International Monetary Fund and the International Energy Agency pressed the Russian government to remove an export tariff from crude oil. They argued that the tariff depressed demand for exports and increased domestic consumption. Nothing happened to either exports or domestic prices after the export tariff was removed. Domestic prices were depressed by low demand, and exports were limited by infrastructure: All the pipelines and ports were already jammed.

Russian government officials pointed to the fall in oil prices partially to explain their failed defense of the ruble and the disastrous government default and currency devaluation of August 1998. The international price of crude oil in mid-1996 hovered between $22 and $24 per barrel, but by March 1998, the price had fallen all the way back to $14. Russian export profit per barrel fell from $4.14 to $0.29 per barrel, despite elimination of the $3.57 export duty. Russia's largest oil company, Lukoil, reported a loss of $1.29 per barrel exported in 1998 through Novorossiysk. Selling on the domestic Russian market was unappealing to the oil companies because less than one-fifth of domestic sales were conducted in cash, and those at deep discount.[27] Devaluation further eroded the ruble value of Russian oil exports.

The Caspian

Caspian Sea oil reserve estimates were revised upward in 1997 to 178 billion barrels of "proven and possible reserves."[28] The supergiant Tengiz field in the northeast of the Caspian holds 6–9 billion barrels, according to oil industry estimates.[29] Tengiz was the largest oil find in the last quarter of the twentieth century, until the Kashagan formation was explored and indicated to have "six Tengizes in there."[30] Each one would be enough to supply the United States with petroleum for a year, or China for five years at 1996 consumption levels. Major players for this oil, in addition to Kazakhstan, Azerbaijan, and Turkmenistan, are western oil companies and their representative governments. Complicating matters are relations among these parties and, in alphabetical order, Afghanistan, Armenia, Chechnya, China, environmentalists, Georgia, India, Iran, Kurds in Turkey, Russia, and Turkey. Western oil companies were brought in to develop Tengiz in 1990, before the collapse of the Soviet Union, because it lay 900 meters below a salt dome and required western technology to be exploited.[31]

There are two problems with Caspian oil: No one agrees on who owns it or how best to transport it to market. The legal status of the Caspian long remained subject to international dispute, with Russia and Iran and sometimes Kazakhstan maintaining that it is a lake, and Azerbaijan, Turkmenistan, and sometimes Kazakhstan—backed by the United States—asserting that it is a sea. Under the Law of the Sea, countries along the water would divide up resources down the middle, that is, equidistant from the boundaries along the shores of the sea. But if it is a lake, each surrounding country would own out only a few kilometers into the middle of the body of water, and all the countries would own the resources in between in condominium. The Russian plan for the Caspian's division has been called the "doughnut" approach because it would leave a central hole in the center where no one country owned the oil resources.[32] That is, Russia and Iran would get bigger shares if the Caspian is a lake, and they would be able to assert control over production and transportation systems. Azerbaijan, especially, would lose significantly under that definition. Iran especially would gain because there appear to be few resources immediately off the coast of Iran. Objectively, the body of water is closed and appears to be a lake, but law and logic are different subjects.

In the Tengiz deal, Chevron paid $1.1 billion for half of a joint venture with rights to develop the oil, a company called Tengizchevroil.[33] The agreement has been expanded to $20 billion of development involving the Kazakh government, Chevron, Mobil, Lukoil of Russia, and others. Another deal to develop the Karachaganak field in the northwestern corner of Kazakhstan, near Orenburg, was worth $8 billion and probably involves 2 billion barrels of oil. Offshore Caspian resources—as opposed to reserves—may amount to 60 billion barrels. This may someday put Kazakhstan in the same category with Russia in reserves and in the same category as the fabulously rich Middle Eastern oil nations. In 1999, Kazakhstan's population was only 17 million people, enjoying per capita GDP of only slightly over $1,000. Oil exports in 1995 totaled only 74,000 barrels of oil per day, worth about $25 per capita.[34] That amount approximated the quota Kazahkstan was allowed on Russia's oil export pipeline system. The export transportation constraint, and the minefields across which any new pipelines would go, create a whole new aspect of geopolitics that some find intoxicating and many find dangerous.

The Caspian Pipeline Consortium (CPC) reached agreement in late 1998 to build a pipeline that the Russian and Kazakh governments approved in 1996. The line would transport Tengiz oil from Kazakhstan across Russia to Novorossiysk, where it would be off-loaded into tankers for a journey across the Black Sea and down the narrow—some say perilous—Bosporus Straits. It would span over 1,500 kilometers (940 miles), cost $2.3 billion, and carry as much as 1.3 million barrels per day by 2014. This pipeline has been alternately viewed as complementary to and competition for pipelines from the southern Caspian to Turkey. The latter would not be directly controlled by Russia, a major advantage for western companies, whom the Russian government keeps at arm's length to protect its own industry. Second, it is closer to markets—as the crow flies—because it lies between the Mediterranean and Siberia. The CPC line will be vulnerable not only to Russian political control, because it would cross Russian territory, but also to control by the Turkish government, which controls tanker traffic through the Bosporus Straits.[35]

Near the other end of the Caspian, the "contract of the century" involves a consortium called the Azerbaijan International Operating Company (AIOC), which includes the Azerbaijani national oil company, British Petroleum, Amoco (now merged with BP), Exxon, Unocal, Pennzoil, Lukoil, and five other companies.[36] Their payment was the opportunity to choose 10 percent of the area identified as promising for oil production for their own drilling.[37] The AIOC plans to develop three fields with an estimated 4 billion barrels at a cost of at least $8 billion. The consortium produced some 75,000 barrels of oil per day in 1998, which it exported north from Azerbaijan through Chechnya to Novorossiysk on the Black Sea. The pipeline had been built to carry products in the opposite direction but was never used. The consortium modified and completed that line while working to rebuild and connect a products pipeline west across Georgia to Supsa, also on the Black Sea. When completed, the two lines would provide export potential of 200,000 barrels of oil per day. Controversy has surfaced over the choice of route for a main export pipeline capable of carrying up to 1.5 million barrels per day. The U.S. government argues for a line west then south across Turkey to the Mediterranean, but oil companies and regional governments argue that a line across Iran would be cheaper and more secure. Environmental concerns also drive the debate because exporting large volumes across the Black Sea and through the narrow, winding Strait of Bosporus risks a catastrophic spill.

Turkmenistan is the third major prize in the Caspian. The nation is controlled by the head of the Democratic Party of Turkmenistan, the only party allowed. Oil analyst Robert Ebel calls personal contacts the only real currency in Turkmenistan.[38] Turkmenistan's population numbers only about 4 million, and its GDP only about $4 billion.[39] The U.S. Department of Energy employs many fewer people than the population of Turkmenistan but has a budget several times higher that Turkmenistan's GDP. Turkmenistan possesses approximately 165 exajoules of gas, enough to provide, for example, all of Ukraine's energy needs for roughly three decades. The most promising part of the Caspian sea bed may be Turkmen, and it is untouched.

Neighboring Uzbekistan has a population of about 22 million, with a GDP per capita of less than $1,000. More populous than any of the other Central Asian states, it has historic cities—Tashkent, Samarkand—and a strong president, but significant development problems. The nation has little oil to export, perhaps some gas, but it has poor access to markets. Its infant mortality rate, an important indicator of development or the lack thereof, is 3.4 percent, not quite twice that in Russia but nearly five times the U.S. average.[40]

Despite large resource numbers, the future of the Caspian is highly uncertain. The region will be of interest for the long term, but when companies boast of a "$20 billion joint venture," such as the Chevron-Kazakh Tengiz deal, they are describing plans that stretch over forty years. With oil at $15 per barrel, the economics of off-shore oil production and export pipeline construction become dubious. Caspian offshore oil is more expensive than Middle Eastern or even South American oil, and low oil prices could make Caspian oil unprofitable.[41] The AIOC argues that its costs compare favorably to North Sea offshore platforms where its member companies made profits.[42]

The inherent risk due to conflict and geopolitics may make such a venture less than worthwhile, and output may long total less than 2 million barrels per day, or less than the production of Canada, Nigeria, or Kuwait. Moreover, it may be impossible for the United States, or any other outsiders, to project military force into the region. It will not be possible to send U.S. gunboats or aircraft carriers to the Caspian. Airlifts over Iran, Russia, Iraq, and Afghanistan might be difficult. A "Desert Storm" could be mounted in Azerbaijan only with great difficulty and expense. Cooperation, as opposed to gunboat diplomacy, may be the only chance for Caspian energy development.

There are some signs of increased cooperation in the Caspian area. Russia's Lukoil company, after all, is an important partner in the AIOC. And in an agreement that seemed to abandon Russia's long-held position that the Caspian is a lake, not a sea, President Yeltsin signed a bilateral agreement in mid—1998 with President Nursultan Nazarbayev giving Kazakhstan the right to develop oil along the Kazakh shore in the Kashagan formation in the northeastern end of the Caspian. This agreement was billed as the first that defined the legal status of the Caspian after the demise of the Soviet Union, and it is to be a model for other states. Critics of the agreement noted that Kazakhstan had not won the right to build undersea pipelines, among other things, and that Russia could still control such investment decisions. Kazakhstan and Russia may have cooperated in part due to the friendship of Presidents Yeltsin and Nazarbayev, and in part because the population of Kazakhstan is ethnically one-third Russian and only about half Kazakh. Russia, for its part, argued that seismic activity in the region made underwater pipelines dangerous environmental hazards.[43]

Even if optimistic projections are realized, the Caspian area would supply only about 3 percent of the global oil and would achieve roughly the importance of the United Kingdom or Norway. The Caspian thus might be considered a new North Sea, but it is unlikely to be a new Middle East, at least not for many years.[44]

Oil's export pipeline problems converge with those of natural gas. Moreover, the southern Caspian eventually will yield more gas than oil. But oil has one other, special issue, and that is refining of products for domestic use. Natural gas, which is mainly produced separately from oil, carries a different portfolio of issues.

Refining

Refining of oil into useful products such as jet fuel, diesel fuel, heating oil, and gasoline suffered from a fundamental flaw in Soviet production, an emphasis on quantity over quality. At the time of its demise, the Soviet government directly controlled capacity to refine 3 billion barrels of oil per year (8.4 million barrels per day, or 18.4 exajoules).[45] A large share of refinery capacity in the 1950s was located in the Volga-Urals region, but with development of the crude oil pipeline system, refineries were developed closer to markets because crude transported by pipe costs one-half to two-thirds less to transport than product transported by rail.[46] Refining centers were developed in Ukraine, Belarus, Lithuania, around the Moscow region, at Omsk, and in the Russian far east. A disproportionate share of refining remained in the Volga-Urals region, nevertheless, taking oil from pipelines flowing west.

Russia's twenty-eight refineries have a capacity to refine 4 million barrels of oil per day, roughly 8 exajoules. An astounding 38 percent of their output is heavy residual oil—mazut—that is fit mainly for burning in power plants. Advanced refining technologies such as catalytic cracking and hydro cracking are used far less in Russian than in the U.S. oil industry, although that is partly due to the higher demand for gasoline in the United States (see Table 4.1). Unleaded gasoline accounts for only one-fifth of the total produced. Some $8 billion in investment is needed by the year 2000 to upgrade Russian refineries.[47]

Three dilemmas confront the refining industry. The first is collapsing demand. In the Russian Federation, oil consumption fell by 50 percent from 1990 to 1996. Even in 1996, consumption—as opposed to oil production—was still falling, dropping by one-eighth compared to 1995. Ukrainian demand over the same period fell by 75 percent. By comparison, South Korea and Taiwan doubled consumption during that time. The world as a whole increased oil consumption by 6 percent, despite the drop in the former Soviet Union. Indeed, 1997 was a record year for oil production and consumption worldwide. Eastern European countries' demand declined, but much less, dropping 10 percent in the Czech Republic and 25 percent in Bulgaria, Hungary, and Romania. Polish de-

TABLE 4.1   Refinery Characteristics

	Post-Soviet	U.S.
Vacuum Distillation	27%	44%
Catalytic Reforming	3%	24%
Catalytic Hydro refining	9%	12%
Catalytic Cracking	5%	34%
Catalytic Hydrocracking	~0%	8%

SOURCE: Office of Technology Assessment, *Fueling Reform: Energy Technologies for the Former East Bloc* (Washington, D.C.: Government Printing Office, 1994), 47.

mand increased 10 percent in 1996, at a slower rate in 1997. Polish oil demand in 1998 was 10 percent higher than 1988 consumption.[48]

If Poland is any guide, demand will not return to pre-revolutionary levels until a decade after reform was initiated.[49] But Poland never fell more than 20 percent in oil demand, and countries where demand fell twice as far will take longer for demand to recover. Need for refining capacity is not likely to exceed the 1990 level in Russia and throughout the region until at least 2005 and probably much later. Serious over-capacity is likely to remain a problem in Russia and Ukraine for some time, particularly in the Volga-Urals region and the North Caucasus, where capacity is most over-built relative to regional demand, making the elimination of some refineries inevitable.[50] Upgrading surviving refineries will be a costly, slow process.

A second dilemma, however, is lack of capacity where it is needed. Local shortages have been so severe in the far east that the use of portable refineries has been considered.[51] Refinery development projects are being promoted in Tyumen, Tomsk, Omsk, and the Far East.

A third dilemma may be the most difficult: the technologically backward state of the region's refineries. The average output of Soviet-style refineries is, as noted, almost 40 percent residual oil, compared to 6 percent in the west.[52] It is the lighter products—gasoline, diesel, heating oil—that are the objects of refining. Although Russia will likely never need to produce the extremely high ratio of gasoline per barrel of the car-crazed United States, expensive capital upgrades to Russian refineries will be needed.

Lighter oil products can be produced by adding distillation columns to boil off more of the lighter product. Refiners produce light products by placing heat under crude oil in distillation columns. Lighter compounds such as propane, butane, and gasoline have lower boiling points and boil off first and can be separated, collected, and distilled. Heavier products such as kerosene are next to boil. The lighter products have higher hydrogen to carbon ratios, and the hydrogen bonds contain the energy sought in burning fossil fuels. But to make lighter products out of the longer-chained, less elegant compounds requires catalysts to crack and reassemble molecules. A catalytic unit upgrade to a typical Russian refinery would cost on the order of a billion dollars.

Ukraine has refinery capacity of 1.2 million barrels of oil per day.[53] Of these, Lysychansk and Kremenchuk are only twenty years old and reasonably competitive. Lysychansk, in fact, has been substantially upgraded with modern control systems, and both refineries have plans for adding distillation units or catalytic crackers. However, only Lysychansk of all Ukrainian refineries in 1996 operated at more than 50 percent capacity. The rest ranged from 8 to 38 percent.[54] Operation was often possible only by tolling, getting paid a fee for processing oil lent by the owner. That practice, however, means that the refineries do not capture the full value added of their work, and therefore reduce even further their profits. The practice stems from the refining companies having too little cash to buy crude oil.

Russian Gas

Natural gas production in the Russian Republic stands in marked contrast to oil's collapse. Gas output remained steady throughout the 1990s, declining only in 1997 but rebounding in 1998 (see Figure 4.4). Gas differs from oil because it is less technologically demanding and because Russia maintained investment in development of the resource. In energy terms, Russia produces three-quarters more gas than oil.[55] At 21 exajoules (561 billion cubic meters) of production, only the U.S. gas industry is Russia's peer. The annual wholesale value of Russia's gas production has reached an estimated $75 billion, with exports worth $15 billion at market prices.

Russia's gas industry creates opportunities for low-carbon and relatively clean energy for Europe, Asia, and Russia, although it will increase pressure on the world's largest and most fragile wilderness. Significantly, this potential is controlled by a single company, Gazprom, which controls more revenues than most governments. Gazprom in 1996 ranked as one of the world's ten most profitable energy companies. It had sales worth over $20 billion and property and equipment worth over $44 billion, and

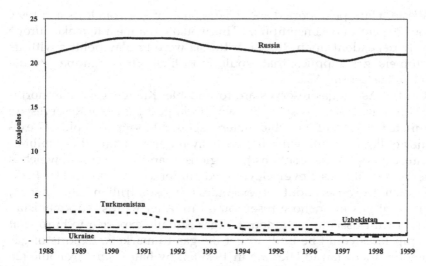

Figure 4.4 Natural Gas Production, Selected Transition Economies, 1988–1999.
SOURCE: BPAmoco, *Statistical Review of World Energy, 1999,*
http://www.bpamoco.com/worldenergy/; author's estimates.

in 1997 paid taxes and late-payment penalties of $3.1 billion.[56] Gazprom employed some 350,000 workers, including those providing social services, although it has moved to organize the latter into a separate division, ultimately to be sold off.[57] The gas company produces three and one-half times more energy than Exxon.[58] Proven gas reserves total over 1,600 exajoules, with an additional 1,600 exajoules likely to be developed: a 100–200 year supply of gas for Russia at current levels of consumption.[59]

Natural gas production technology is comparatively simple. Although the exploration techniques are similar to those for oil, production essentially amounts to placing a straw in the ground and allowing the gas to flow into a pipeline. Gas must be cleaned of any sulfur, liquids, and contaminants at the well-head, and cleaning can sometimes be locally hazardous. Some Kazakh and Russian gas contains as much as 75 percent sulfide, but after cleaning and transport to market it is ready to use. It is, in fact, the transportation of the gas that is the more challenging problem because gas transmission lines are pressurized at several tens of atmospheres. Long-distance transmission lines require compressor stations at intervals of 80–160 kilometers (50–100 miles), such as are found all along the length of the pipeline from Urengoy or Yamal to Europe.[60] Valves must be provided at distances of 10–30 kilometers for releasing gas in case of compressor failure or other problems.

Western Europe imported about 5 exajoules of gas from Russia in 1999, about 20 percent of consumption. This would not appear to make Europe overly dependent upon Russia. If Russia were to play politics with its customers' gas supplies, that would most likely drive Europe into the arms of the Central Asians.

Central Asia's gas reserves are formidable. Karachaganak, in northwestern Kazakhstan, may be the world's largest gas field. Reserves are estimated at 1.3 trillion cubic meters (about 50 exajoules), plus 37 exajoules of liquids. Producing the gas may require advanced capabilities because it apparently contains prodigious quantities of sulfur, which is one reason it was never developed under Soviet control.[61] Turkmenistan possesses about 110 exajoules (almost 3 trillion cubic meters) of natural gas in onshore reservoirs.[62] In other words, Turkmenistan could rival Russia in the amount of gas actually exported to Europe, if the pipelines could be built to get the gas to markets. Exports of gas from Turkmenistan to Ukraine in 1993 fell by half from 1 exajoule (26 billion cubic meters) due to nonpayment, a problem the exporter continued to experience with Ukraine to the end of the decade.[63] A joint stock company called Turkmenrosgaz was formed in 1995, with the Turkmen government controlling 51 percent and Russia's Gazprom 44 percent. Some competitors worry that this deal makes Gazprom a player in the Caspian and will further undermine the region's independence from Russia.[64]

Pipeline Geopolitics

Pipelines and proposed pipelines trace the energy geopolitics of the former Soviet Union. Pipelines are essential because the oil and gas wealth lies far from markets—primarily in Western Siberia, the Caspian, and Eastern Siberia—and export routes by water to Europe or Asia do not, for all practical purposes, exist. The oil resource lying under the Caspian is prodigious, worth perhaps $3 trillion, more, with rebounding prices. Half the world's natural gas—as well as oil resources exceeding that of the Caspian—lie under the alternately frozen and flooded wilderness of Siberia, far from both Europe and China.

Key energy trade issues include the fact that the Russian pipeline system is not matched well to current markets; that Russia is a formidable negotiator when it comes to pipeline routes and the control of resources it wants; that a half-dozen brutal conflicts surround the Caspian; and that the most desirable non-Russian routes would traverse Iran, which the United States considers a pariah because of its export of terrorism and development of weapons of mass destruction.

Russian oil moves by pipeline: 95 percent of it. The oil pipeline system is a monopoly controlled and operated by Transneft, nominally a joint-stock company, and by Transnefteprodukt. The former handles crude and the latter carries products. Russia's oil pipeline system overall extends 50,000 kilometers (31,000 miles) and can carry 13 million barrels per day. A regulatory commission now, in theory, controls the operations of these two pipeline monopolies.

Russia in 1996 exported 3.8 million barrels of oil per day, almost two-thirds of its total production. Export policy earlier in the decade affected prices by restricting exports; products flooded the domestic market and dropped prices to 29 percent of world level in 1994.[65] Russia's major export pipeline is the Druzhba, or "Friendship," which can carry up to 1.2 million barrels per day to Eastern and Central Europe. Due to declining demand, however, this pipeline has carried only about 60 percent of that capacity.[66] Declining oil demand in Eastern Europe has meant a scramble to export oil through ports, which, along with the pipelines to them, have become bottlenecks. Export capacity to seaports for European markets, including the pipeline between Tikhoretsk and Novorossiysk on the Black Sea, has been operating at maximum capacity. The port at Novorossiysk recently handled almost a third of exports to countries outside the old Soviet sphere, or about 600,000 barrels per day. Odessa, now Ukrainian, handled about 210,000 barrels per day of Russian oil exports. Odessa, however, is interested in moving oil in the opposite direction by building port facilities capable of importing oil from non-Russian sources brought by ship across the Black Sea.[67] Such a move would increase Ukraine's energy options and reduce its perceived dependence on Russian oil.

The Caspian has become the focus of international political attention, not just because of its potential wealth but because political, geographical, and technical problems complicate the development and use of export pipelines. These complications increase tensions and raise the stakes far beyond those of cheap gasoline. The first oil pipeline—first in the world as well as the region—was built from Baku on the Caspian to Batumi on the Black Sea in 1897–1905, about the time Stalin was expelled from a Georgian seminary school. Today, as pipeline routes are plotted out of the Caspian in every direction, Stalin could find revolutionary company, also in every direction. Conflicts, both ethnic and ideological, surround the Caspian and thwart efforts to take oil and gas to European or Asian markets. Russia and Chechnya continue a bitter war in the northwestern Caspian region, and the conflict halted the flow of oil through a pipeline passing from Baku to Novorossiysk through Dagestan.[68] Georgia, devastated by civil war in the early 1990s, lies to the west across a major alternative export route to the Black Sea. Its president, Eduard Shevardnadze, narrowly escaped assassins' rocket-launched grenades in early 1998. The desperation for energy that can be caused by war was demonstrated in Georgia during the early post-Soviet years. An oil pipeline from Baku was drilled by thieves for oil stored in the shut-down line and refined in makeshift refineries to produce low quality diesel and heating oil for a desperate population in winter.[69] Georgia's prospects for stability improved, however, with peaceful parliamentary elections in 1999 that seemed to embrace reform.

To the southwest, 20,000 people died in the Azeri-Armenian conflict in the early 1990s. Although fighting stopped in 1994, hundreds of thousands of Azeri and Armenian refugees remain, many of them living in rail cars, and Armenia still occupies one-fifth of what was Azerbaijan.[70] Iran occupies the southern end of the Caspian. Turkmenistan to the southeast is relatively stable, due to its authoritarianism. Kazakhstan to the northeast stands as a relatively stable country, if not a democratic one. Mountains on the east and the vast distance to Chinese markets make Asia, at best, a long-term export route, although some investors are considering it.[71]

The Caspian export problem is epitomized by the dilemma faced by Kazakhstan. Kazakhstan's quota on the Russian oil pipeline to Novorossiysk is only 76,000 barrels per day, less than one-third of current production in the giant Tenghiz field. Some 40 percent of Tenghiz oil is now moving by means of rail or ship.

The Azerbaijan-Russian pipeline runs almost 1,400 kilometers (850 miles), cost $50 million, and until the second Chechnian war, carried 0.1 million barrels per day. A first pipeline through non-Russian territory, the Azerbaijan-Georgia pipeline, crosses 880 kilometers (550 miles), cost $315 million, and can carry 0.1–0.2 million barrels per day.[72] Another Azerbaijan line, perhaps ultimately across Turkey, is being discussed and debated, and would possibly carry 800,000 barrels per day.[73] Total foreseeable oil export capacity would thus seem to be limited to under 2.5 million barrels per day, less than the production of Iran although almost as much as the United Arab Emirates.

Kazakhstan and Azerbaijan reportedly signed a deal in June 1997 to build a pipeline from western Kazakhstan, first to Turkmenbashi in Turkmenistan, then under the Caspian to Azerbaijan.[74] The problem Turkmenistan faces—or rather that the multitude of foreign companies developing Turkmen gas face—is a chess game in both a physical and strategic sense. Turkmen gas could move queen-like straight ahead to the objective, western Europe, if it could send its oil across Russia in existing Russian or new pipelines. But Russia's gas pipelines are more or less filled and are in any event controlled by Russian actors. Of course, any new Russian pipeline would also be controlled by Russia, and possibly Ukraine or Belarus as well. Gas could be taken south rook-like across

Iran to the Persian Gulf and then as liquefied natural gas by tanker to Europe, but the Iranian route presents political problems because the United States opposes dealing with Iran. Gas could be taken west, knight-like, first around war zones in Nagorno-Karabakh (between Azerbaijan and Armenia) and Chechnya (between the Caspian and Georgia), and then across Georgia to the Caspian or down through Turkey to the Mediterranean. It would be financially feasible to build a gas pipeline south from Turkmenistan across Afghanistan to Asia, for domestic power generation and international marketing, but the lack of any central authority in Afghanistan makes this a dubious choice.[75] When the U.S. government struck a terrorist site in Afghanistan in August 1998 in retaliation for bombings of U.S. embassies in Nairobi and Dar-es-Salaam, it affected a deal between Unocal to build a natural gas pipeline across Afghanistan from Turkmenistan to Pakistan. The project, which had been scheduled to start construction in 1998, was suspended indefinitely.[76]

Still another Caspian export route would send Turkmen gas across Afghanistan to Pakistan and India. Several oil companies are promoting two different projects for that purpose.[77] An option favored by the U.S. government would build undersea pipelines to carry Turkmen and Uzbek oil and gas across the Caspian to link up with pipelines in an east-west corridor.[78]

A highly-touted alternative is an Azerbaijan route west through Georgia, then southwest across Turkey to Ceyhan on the Mediterranean. From the Azeri and oil company perspective, this route would have the political advantages of avoiding Russia and Iran, both allies of its enemy, Armenia, as well as adversaries in the control of Caspian oil. It would have the disadvantage of proximity to four zones of conflict: between Armenians and Azeris, Georgians and Abkhazians, Russians and Chechnians, and Turks and Kurds.[79] Western oil companies reluctantly endorsed this pipeline in late 1999.

To the southwest of the Caspian, toward the European markets, Armenia has been cut off from the east by Azerbaijan and from the north by the war in Chechnya. After fighting with Azerbaijan over Nagorno-Karabakh, the Armenian enclave, Azerbaijan imposed a punishing oil embargo. Armenia survived in part on donated kerosene from the United States, but became desperate enough for energy to restart a Soviet nuclear power plant so dangerous that even Minatom, not renowned for its safety record, had closed due to its location on a geological fault. At one point, Armenia paid for gas from Turkmenistan with jewelry. The desperate Armenian-Azeri conflict, although recently dormant, significantly affects regional energy prospects.[80] More feasible politically is a gas pipeline that Russia's Gazprom and Italy's Eni agreed to build under the Black Sea from Dzhugba to Samsun, connecting to an existing line to

Ankara. However, this line would be 30 percent deeper than any existing gas pipeline and run through corrosive hydrogen sulfide gas contained in the Black Sea waters.[81]

A visionary proposal would take gas and oil east over the mountains to China and Japan. The distance of 5,800 kilometers (3,600 miles) would make it the longest pipeline in the world, and it could cost $15–25 billion. Nevertheless, the Chinese National Petroleum Company (CNPC) bought rights to develop oil and gas resources in the Aktyubinsk field, receiving a 60 percent stake for $4.3 billion and a pledge of $3.5 billion for construction of a natural gas pipeline to Xinjiang province in western China.[82] This project has raised eyebrows because gas fields in Eastern Siberia are closer to major Chinese markets and would thus have lower transportation costs. A pipeline would originate 350 km north of Irkutsk at Kovitinskoye and cost $5–7 billion. Rusia Petroleum, a subsidiary of the Russian firm Sidanko, holds the license to develop Kovytinsk, and in late 1997 concluded a joint venture with British Petroleum (BP) to develop 20 billion cubic meters of gas per year for export to China.[83] Bilateral discussions between President Yeltsin and Premier Jiang have included positive discussions of this Russian-Chinese gas deal.[84] Failure to pay taxes and, some would say, domestic Russian efforts to seize control of Sidanko by questionable means, had by late 1999 put this deal in jeopardy, along with BP's half-billion dollar investment in the Russian oil company, although a compromise may have been brokered in early 2000. This situation is often cited as giving credence to cynics' view that the Russian oil sector is corrupt and worthy only of the most speculative type of investment.[85]

Most of the initial solutions to gas and oil exports from the region have involved accommodation. That means the west and Central Asia have made Russia a partner on both resource and pipeline deals.[86] It has also prompted a slight U.S. warming to Iran because Iran offers an alternative to Russian export routes. This prospect is complicated further by Turkey's desire for the gas, because Turkey is a member of NATO and considered critical to U.S. interests in the region.[87] Some observers point to Iran's brutal suppression of student demonstrations in 1999 as indicating that Iran is unlikely to liberalize in the near future.[88]

It has been argued that Kazakhstan has attracted foreign investment in its oil sector because the nation is run dictatorially rather than democratically. The argument is that it has been easier for President Nursultan Nazarbayev to enact laws favorable to foreign investors than former President Boris Yeltsin because the latter must obtain the consent of a nationalistic and xenophobic legislature.[89] This theory would also apply to other major players, including Azerbaijan and Turkmenistan, but not to Iran. However, there is probably another reason why the Central Asian nations find themselves favored by western companies and their governments over Russia: Russia is far more difficult to control than the smaller states and has less in relative terms to gain by dealing with foreign interests. Kazakhstan's hopes of increasing per capita income from $1,000 annually to a higher number by developing its oil wealth is more plausible because it has a small population. Russia's much larger population—many of whom remember their superpower status—have a different point of view. A question for U.S. and other western policy makers is whether cutting deals with dictators speeds or retards the day when democratic reforms come to the Caspian, and whether it matters. Another concern, closer to the American heart, is whether U.S. soldiers would have to do for Kazakhstan what they did for Kuwait, and, given the fact that the Caspian is landlocked, except via the Volga through Russia, whether projecting U.S. force into the region would even be possible.

The sharpest conflict of interest for U.S. policy makers promoting oil and gas development in the Caspian comes with Iran. Ostracism of Iran has been part of U.S. foreign policy for two decades, justified by the effort to prevent Iran from gaining atomic or biological weapons and from visiting terrorist attacks on U.S. interests and those of its allies. But softening of this line may come by permitting Turkmenistan to export gas across Iran to Turkey. The choice of Iran over Russia would seem an odd one, but is evidence of the power of energy development. Efforts to build a $1.6 billion gas pipeline across Iran have been rumored to be acceptable to the United States to permit Turkmenistan to avoid having to deal with Russia for pipeline rights.[90] The first leg of a gas pipeline connecting Turkmenistani production with northeastern Iran began operation in late 1997. Iran financed most of the pipeline construction costs, and will repay these in gas over three years.[91] Iran and the Caspian oil prize have so impressed the United States that it created a position on the National Security Council of special assistant to the President for Caspian Basin Energy Diplomacy.[92]

Another conflict that U.S. policy makers will have to resolve is climate. Natural gas produces 30 percent less carbon dioxide per unit of energy than does oil, and 50 percent less than coal. Harmonization of European, Japanese, and U.S. climate and trade policies would require a preference for gas projects. Currently, no mechanism exists for bringing about such a harmonization, and the relevant trade agencies, especially the U.S. Export-Import Bank, concede that their mission is trade first, environment second.

The Europol Yamal pipeline will carry Russian gas some 4,200 kilometers (2,500 miles)[93] from Russia's Yamal Peninsula across Poland to Western Europe. At a cost of $24 billion, it will provide gas as a substitute for dirty Eastern and Western European coal, which produces twice as much

carbon per unit of energy as gas. There is a straightforward energy reason, however, to shift more attention to gas. For example, the Caspian
Pipeline Consortium has been the focus of much government attention,
but it will carry only half the energy to be carried by the Europol natural
gas pipeline.

Although one might think that U.S. commitments in the Kyoto Protocol on curbing greenhouse gas emissions would affect U.S. promotion of
oil development, there is no evidence to that effect. A higher—but un-
stated—priority is reducing Russian control over Central Asian energy
exports. In this matter, U.S. government and private interests converge.
The race to control the resource by controlling export routes has been evident in competing offers to finance pipeline construction. Iran offered to
finance a pipeline beginning at the Iranian port of Neka at the southeastern extreme of the Caspian. The tender offer of $400 million for a 392-km
oil line to Tehran was to be the first leg in a pipeline to Kharg Island off
the south coast of Iran in the Persian Gulf. Such a pipeline would be the
cheapest and least environmentally risky route out of the Caspian region
because it would avoid the dangerous Bosporus Straits on the Black
Sea.[94] The route would avoid war zones, including areas of conflict in
Iran. It is backed by Dutch, French, and Russian interests. However, the
United States, supporting Azerbaijan and Turkmenistan, has offered to
provide billions in financing for a Baku-Ceyhan route, for both oil and
gas. Turkmenistan could connect to this outlet via new lines that would
be built under the Caspian.[95] The U.S. Export-Import Bank and the U.S.
Department of Energy have expressed support for this route, with Ex-Im
offering unlimited support. "There is no limit to the amount of money we
will commit," to such projects, Ex-Im president James Harmon has been
quoted as saying.[96]

Unfortunately for U.S. policy, the Iranian route would be considered
a threat to U.S. security. The United States argues that its intention is to
avoid allowing any one nation a stranglehold on Caspian exports. This
policy argues for multiple outlets, especially an east-west corridor
through the Caucasus, while nominally supporting Russian pipeline
development and continuing to oppose Iranian routes. These elements
stem from U.S. interest in supporting the development and independence of countries in the Caspian region, diversifying global energy
supplies as a means of enhancing U.S. security, and promoting U.S.
trade.[97]

Reform

Competition is the foundation of market economies. It fosters efficiency
because firms must at least match the value offered by their competitors.

It fosters accountability because firms fail utterly if they fail to compete. Unfortunately, the state energy monopolies of the Soviet era have been replaced by private monopolies. Competition in the energy sector is hard to find anywhere in the region.

Commercialization of the energy industry in the Soviet Union began with the creation of Gazprom in 1989 and was followed in 1991 by the creation of Rosneftegaz. These state-owned but commercialized firms were given responsibility for production and transportation of the fuels, and policy matters were delegated in 1991 to the Russian Ministry of Fuels and Energy. Rosneftegaz was changed in 1993 to Rosneft to manage the state's oil holdings.

The Russian industry has been privatized into vertically integrated joint stock companies, including Lukoil, Surgutneftegaz, Yukos, Slavneft, Sidanko, Vostok, Orenburg, Tyumen, and Rosneft. Lukoil is the largest vertically integrated company, producing about one-seventh of Russian oil in the mid-1990s. Lukoil and Yukos have been called Russia's Exxon and Mobil, respectively, and Lukoil president Vagit Alekperov has been compared to John D. Rockefeller.[98]

The Russian government has created a series of regional oil monopolies. The main function of voucher privatization has been to transfer oil industry control to private hands, but without injecting much capital. Each of the major new producers dominates regional production to the virtual exclusion of others. Rosneft dominates European Russia, including Kaliningrad, Komi, the North Caucasus, and the Volga region. But whether one company dominated Texas and another Oklahoma would probably matter little to New York, as long as there was competition for the latter's market. That, more than anything else, depends on access to transportation: pipelines and rail. Russia's pipeline system is already managed as a common carrier by Transneft. One key question is whether a regulatory regime can be made sufficiently powerful, competent, and accountable to ensure fair, competitive access. Another key question is whether local competition for wholesale and retail sales will be possible. If the vertically integrated companies control all the distribution nodes in a region, it will not be, and market reform will have failed.

Some forty western oil joint ventures operated in Russia at the end of 1996. Joint ventures produced 154 million barrels of oil in 1996 and accounted for almost 8 percent of Russian output.[99] The Russian government retained only 17 percent of Lukoil, and the U.S. firm ARCO has taken a large stake in it. ARCO has purchased 6.3 percent for $250 million, a deal completed in August 1996. In September 1996, ARCO and Lukoil formed LUKArco, with 46 and 54 percent ownership, respectively, and will together invest $6 billion over the next eighteen years, including construction of the Kazakhstan Tenghiz pipeline. Through a series of deals, Lukoil has interests in the Caspian region. Lukoil, incidentally, has made a nominal investment in the United States, recently opening a gasoline station in Virginia.[100]

Foreign joint ventures, which have occurred mainly in the Timan-Pechora region in the arctic and off Sakhalin, have produced a relatively small amount of new oil. Conoco's Polar Lights project in Timan-Pechora accounted for 70 percent of foreign direct investment in Russia's oil sector in 1995, but it produced only a few tens of thousands of barrels per day. By contrast, most new Russian ventures are in Western Siberia, raising criticism that Russia is saving the best oil prospects and almost all gas prospects for Russian companies. One major controversy erupted when one-third of Yukos was bought by what the U.S. embassy in Moscow characterized as "a previously unknown subsidiary of Russia's Minatep Bank."[101]

The Russian oil sector has been a disappointment to both domestic and foreign interests. A combination of low prices and high taxes made it unprofitable to increase oil production in the mid-1990s. The economic crisis hit the oil industry hard, with Russia allowing the number of idle wells to more than quadruple, from 7,000 in 1988 to 32,000 in 1993. Being idle may have destroyed or caused great damage to the wells. The Russian government allowed western firms into Western Siberia in the early 1990s to rehabilitate those wells, which they did at a reported cost of over $70 million per well.[102] Early in 1998, analysts forecast that the oil price decline would hurt Russia's economic recovery and would cost the government $3.5 billion in lost tax revenues.[103]

Increasing output will require better technology, economics, and environmental protection. New drilling methods both reduce cost and environmental impacts. The "slim hole" technique uses a smaller-diameter drill bit and pipe and so takes less capital, energy, time, and 25–40 percent less cost to do an exploratory well. Because it is removing less earth and using less drilling mud, there is less waste product for disposal.[104] In Western Siberia, Amoco has entered a deal with Yuganskneftegaz and Yugraneft that gives the company access to some of the 5 billion barrels estimated in the fields. Amoco, in return, would provide access to horizontal drilling technology. The deal could be worth several tens of billions of dollars.

The tax structure of the oil and gas industry in Russia has been criticized because the sector carries a tax load below that in other energy-producing countries. Moreover, only about half of all energy taxes required by law were collected in 1996.[105] The largest share of unpaid taxes came from the energy sector, and the government went to the length of putting tax collectors and auditors onsite in energy enterprises.[106]

Outside of Russia and the Caspian, petroleum sector reform has been slow. Despite its reputation as a leader in reform, Poland has been slow to privatize the energy sector, especially oil and gas. Poland in 1996 passed the "Law on Commercialization and Privatization of State Owned Enterprises" after three years of debate. This law by 1999 was beginning to move privatization of the refineries, as well as the production and distribution facilities. Polish reform has been slowed in part by the usual concerns about control and windfall profit, but also by concerns about energy security.

Poland produces no oil, but its major refineries are connected to Russian oil suppliers by pipeline. Petrochemia in Płock is connected to the Druzba pipeline, while Gdanska in Gdansk is connected to Petrochemia by a line called Pomorski. Gdanska has also imported oil by sea from the Middle East, including Iraq, and the North Sea.[107] The two provide 90 percent of Poland's refining capacity, Płock being the third largest refinery in Central Europe. With almost all of Polish high-grade energy—which is to say oil and gas—coming from Russia, and with Polish-Russian relations over time being less friendly than indicated by the name of the Russian oil pipeline that serves the country, Poland has been intent to seek change that provides alternative options and increases the diversity of supply. Such options do exist, including transport by sea and pipelines from Western Europe, the North Sea, and even Algeria.[108] Poland, like the Czech Republic, would also prefer natural gas to coal and oil for environmental reasons, but because gas and oil are often interchangeable, at least as boiler fuel, additional oil supply capability, could be considered insurance against loss of gas supplies. Poland still imposes oil product quotas to protect its domestic refining industry, and will not fully remove them until at least 1999. Similarly, tariffs on product imports have been extended through the year 2000. The tariffs are 15–25 percent on gasoline, for example, and are higher on Russian production.[109] For Eastern Europe, gas import dependence is a problem with this otherwise attractive supply. Storage and local development of coal bed methane may help, although not completely.

In the Czech Republic petroleum dependence is also a concern. Most of that nation's crude oil is imported using Russia's Druzhba oil pipeline, which passes through Ukraine and Slovakia. All Czech gas comes through pipelines controlled primarily by Gazprom. Transgas, a Czech company, predicts that gas demand will double in that country by 2005. The Czech Republic has diversified its oil supplies by building a new pipeline connecting the Druzhba pipeline to the Transalpine line that originates in the Adriatic at Trieste.[110] However, the state import company Transgas in 1998 signed a $10 billion, fifteen-year deal to satisfy three-quarters of the nation's gas requirements.[111]

Natural gas has seen even less market reform than the oil sector, although the gas industry's performance has on the whole been far better than that of oil. Gazprom, the largest private company in the world, provides 20 percent more energy than Aramco, the Arabian-American oil company, and owns $30 billion in assets. The gas giant produced almost 22 exajoules in 1995, an amount equivalent to one-quarter of U.S. energy consumption, or as much energy as Exxon, Mobil, Chevron, Amoco, Shell, Texaco, and ARCO altogether in their worldwide operations.[112] In the United States, a company with the combined power of the seven leading oil companies would probably be considered a monopoly ready for divestiture.

However, Gazprom's monopoly status is a mixed blessing. The company, out of political necessity, is forced to behave like a state-owned enterprise, subsidizing firms with continued supply despite nonpayment for fuel or with prices dramatically below export value. Thus, Gazprom's revenues in 1994 totaled only $20 billion. Exxon, in contrast, had revenues of $114 billion on sales of 72 percent less energy. Russia's energy economy obviously remains one of allocation and subsidy. Gazprom in 1998 was owed $7 billion dollars for gas supplies, $5 billion of that amount being owed by Russians.[113]

The seven leading Russian oil producers, on the other hand, produce only about 11 exajoules of oil per year, about 85 percent as much oil as their seven U.S. counterparts. The largest, Rosneft, is about half as large as Exxon's worldwide operations, and is comparable in size to Chevron and Amoco. Although Russian oil companies also produce gas—indeed, almost two-thirds as much energy in the form of gas as in oil—they are severely constrained in their ability to develop and market that cleaner fuel form.

These facts have some significance for western energy security. Although Gazprom's size and power threaten Russia's market, Western European importers do have other choices. Russia accounts for half of Western Europe's gas imports, but it provides only about 10 percent of Western European gas consumption. Gas can be stored underground to guard against cutoffs, and many customers can quickly switch to oil, if necessary. British analyst Jonathan Stern argues that the Russian gas "bubble" should even serve to facilitate a competitive market for gas in Europe.[114]

Gazprom's monopoly power extends beyond its dominance of gas production, however, to the all-important control of the gas transmission system. Gazprom has effective control over access to its pipelines, and allows little. That means that neither oil companies producing associated gas nor independent gas companies can export gas or even reach important domestic clients. Ironically, Gazprom is using its financial power to create competition against monopoly distributors in Germany and Romania by building new pipelines and entering into strategic joint ventures. That way, Gazprom could share the higher value of industrial sales and power generated with gas.[115] For twenty-five years, Gazprom has supplied gas to Germany through German companies such as Ruhrgas, but now Gazprom has captured one-eighth of direct sales to Germany through the creation of a joint venture called Wingas, headed by the son of Gazprom's director, Yuri Vyakhirev.[116] This arrangement was permitted under the European Union's new rules liberalizing the energy sector in Europe. It may have signaled much deeper changes to come. Security concerns, however, may be overdrawn even though Russia supplies one-third of German gas. European gas supply is diversified by sources in the Netherlands, Norway, Denmark, and the United Kingdom (see Figure 4.5).

Security—and competition—is more of an issue in Eastern Europe, for both oil and gas. Gazprom's power is more of a problem for Ukraine and Eastern Europe. Gazprom delivers about 2 exajoules of natural gas to Ukraine each year. Energy dependence is one issue driving Ukrainian efforts to sustain a troubled nuclear industry, and is a reason given for that country's reluctance to close the Chernobyl power plant. If Ukraine in 1994 had closed Chernobyl and replaced it with gas-fired power plants, the nation's gas imports from Russia would have increased by 6 percent. Russia has at times reduced gas supplies to Ukraine to punish it for failing to pay for gas consumed. In July 1997, supplies were cut by about one-sixth to force payment.[117] Bulgaria allowed Gazprom to take over full control of the Topenergy company, which handles Russian gas imports.

In Ukraine, oil and gas industry decisions are still made much as before independence. Incentives still flow from cost-plus pricing, state allocation of resources, and state control of management.[118] Ownership of the oil and gas resource in Ukraine remains unclear, as do the laws governing production.[119] The Ukrainian oil industry must pay a 20 percent value-added tax, a royalty that can amount to 20 percent of profits, a profits tax of 30 percent, and a 5 percent sales tax.[120] A $0.75 per barrel transit fee charged to Russia for Druzhba oil piped across Ukraine has caused considerable conflict, including the cut-off in 1996 of oil by Russia (or Ukraine, depending on who tells the story).[121] The nonpayments problem, which affected the entire economy, stymied the expansion of oil and gas production even in joint ventures with foreign firms.[122]

Foreign companies could play an important role in upgrading refineries, from installation of basic controls to catalysis. However, uncertainty with respect to demand makes financing of these upgrades problematic. Refineries have been operating far below capacity—on the order of 60–75

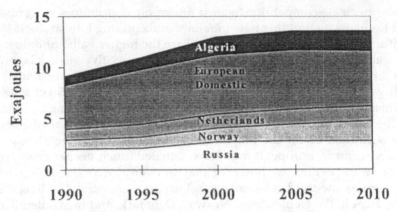

Figure 4.5   Western European Dependence on Natural Gas, by Source.
SOURCE: Cambridge Energy Research Associates.

percent—and their profitability and survival are in doubt. This problem until mid–1995 was greatly exacerbated by the Russian federal government's imposition of export quotas. While that practice succeeded in making domestic oil cheaper, it unintentionally made most of the oil companies unprofitable, thus retarding progress towards markets and efficiency. It is interesting to note that efficiency improvements in a sector with seemingly easy access to hard currency, or at least to revenue streams, are proceeding no more quickly than in the rest of the economy. That may be because the exchange rate is in such great disequilibrium that paying for imported equipment with converted roubles is simply far too costly.

Restructuring Gazprom

Russia's enormous gas industry has been privatized and commercialized but hardly demonopolized. Gazprom is now a joint stock company privatized with a voucher scheme that seems to have been effective at transferring ownership of most shares to private control. Ordinary Russians did get some value out of their shares, enough, say, to trade for an apartment. But a "loans for shares" program soon was needed because voucher privatization raised no capital. This mechanism was not transparent and gave bankers control of vast resources at low prices, discrediting the privatization scheme.[123]

Russia's frustrated young reformers, including former deputy prime minister Boris Nemtsov, vowed to take on Russia's energy monopolies, and had limited success. With Gazprom, he met formidable resistance and was able only to force the organization to privatize its social services and to pay a debt of $1.7 billion in tax penalties owed to the federal government.[124] Gazprom pleaded inability to pay because in 1997 less than 6 percent of gas payments to the giant were in cash, and 55 percent was apparently not paid for at all.[125] There was also the detail of Gazprom having financed President Yeltsin's 1996 campaign. Of course, the reform government had its own problems, being able to collect only two-thirds of targeted revenues.[126] Nemtsov said, "As for Gazprom, despite its status as a huge monopoly, I would not—as strange as it may seem—be in a hurry to start a revolution here, at least as long as its prices are substantially below world prices." Nemtsov borrowed and adapted an old American line about General Motors, saying, "The richer Gazprom is, the richer Russia is."[127]

Chairman Rem Vyakhirev was highly effective at defending Gazprom. He, like former Atomic Energy Minister Victor Mikhailov, showed himself capable of using the nationalist impulse, decrying attempts by foreigners to destroy his company. Vyakhirev warned that, "Anybody who comes to power is going to have to manage to live with Gazprom because without Gazprom, they won't manage at all."[128] The Duma embraced this opinion by voting 332 to 1 against breaking up the company. Former Prime Minister Viktor Chernomyrdin said that, "Natural monopolies, above all the EES [the electric company] and Gazprom, are the backbone of the Russian economy, and we will cherish them as the apple in our eye." *Izvestia* compared Nemtsov's "attack" on Gazprom to hanging a bell around the cat's neck to keep track of it.[129]

Price is a key indicator of reform. Russian gas prices have increased dramatically since reform, but they have not reached levels in comparable markets in the United States or Western Europe. Industry in the United States, for example, paid between $3.50 and $4.00 per GJ in 1997, whereas Russian industry paid an average of about $2.10.[130] Transportation costs were probably lower in Moscow than in much of the U.S. market, but the price did not reflect the opportunity cost of using gas. Nevertheless, pressure by the industrial lobby in Russia forced a reduction in industrial gas prices by 40 percent, although this price was offered only to companies paying cash and in advance.[131]

Managers at Gazprom currently control 35 percent of the state's 40 percent stake in Gazprom. Russian reformers argued that it is wrong that the government control just 5 percent of Gazprom and moved to establish firmer control over the government's share. Gazprom's private owners, especially Oneximbank, which held 3 percent of the company, had already tried to get a candidate elected to the board.[132]

BOX 4.1   GAZPROM FACTS

- Russia's largest company, generating roughly 8 percent of GDP
- Earned almost $23 billion in 1996
- Owns 30 percent of the world's gas reserves–1,200 exajoules
- Russia's largest hard currency earner, earning $8.4 billion in 1995
- Supplies 22 percent of Western Europe's gas.
- Owns transmission pipe enough to wrap around the world nine times
- Earned pre-tax profits of $3.7 billion in 1996
- Owns pipelines worth $25 billion; wells and equipment worth $7.6 billion; machinery and equipment worth $6.1 billion; buildings and roads worth $8.1 billion, for a total of $48 billion

SOURCE: Cambridge Energy Research Associates, "Gazproms' Future: Structure and Strategy" (prepared for the U.S. Department of Energy, June 1995). See also *Oil & Gas Journal* (12 June 1995): 27–32.

One priority is separation of the high pressure gas transmission system from Gazprom's control. Doing so would promote access for independent producers, possibly even non-Russian gas producers in Central Asia. The implications would be cheaper Russian domestic prices, more competition from international producers in the Caspian Basin, and less ability for Gazprom to interfere in national—and international—politics.

Environment

The arctic bio-region is one of special value.[133] It covers 12 percent of Russia, and is linked to terrestrial systems including the deltas of the Lena, Ob, and Yenisey rivers. These areas serve as critical breeding grounds for polar bear, walrus, wild reindeer, red-breasted goose, and polar fox. Oil and gas development threatens these valuable areas. Significant threats to the region include pipelines and roads which fragment habitats, disrupt migration, destroy fisheries, and diminish lichen, which is food for reindeer. The tundra recovers from impact slowly. Prospectors' tire tracks take decades to disappear. Reindeer migrations can be blocked by pipelines. In addition, some of the largest refineries are located in the Arctic. These inefficient operations lack pollution control as well as controls on oil transportation and degrade and destroy wildlife habitat, particularly in the Yamal and Kola peninsulas.

The Western Siberian forest bio-region is of great value because of its large size. Some 50–80 percent of the territory is covered by marsh. The forest is composed of fir, cedar, spruce, birch, and aspen. There, oil development threatens the Siberian crane, beaver, and white-tailed sea eagle. Moose, bear, wolverine, and sable populations are also affected.

Wide areas in the north have been degraded by oil spills, atmospheric releases of toxic metals, and sulfur. In the Tomsk and Tyumen Oblasts, over 90 percent of existing forest reserves have been damaged by industrial activities. These areas lack nature reserves and the protection they afford.[134] The development of Siberian oil is not dramatically different from development of Alaska's. Differences include the greater presence of water in Siberia than in Prudhoe Bay. The environmental impacts of exploration, development, and production have been reduced over the years by improved technology, and have partly been motivated by efforts to cut costs, but not to protect the environment. For example, the footprint of a drilling operation is now somewhat smaller because the industry has implemented cluster drilling operations to reduce the foundation, or drilling pad, area that has to be prepared. Better seismic analysis and horizontal drilling techniques have facilitated this change.

Initial exploration is done by remote sensing—satellites and airplanes monitoring magnetic fields and gravitational variations—or by seismic monitoring stations to detect promising areas. By definition, the local impact is minimal. The first real impact comes when exploratory drilling becomes necessary to determine what remote sensing cannot detect, the presence of substantial oil and gas deposits. The impacts associated with drilling are those of building the roads or the helicopter pads to transport in the requisite equipment and workers, the construction of drilling pads, and the disposal of waste products.[135] Transporting equipment only in the winter months on ice roads will minimize impacts, as will using barges and helicopters in summer.

Drill pads are often gravel lots laid to depths of 1–2 meters (3–6 feet) to support drill rigs, pits, and associated operations. Drilling requires that a lubricating and cleaning material—mud—be pumped down the drill pipe and back up the well between the drill pipe and the cement that lines the well hole. This mud lubricates both the turning pipe—in rotary drilling—and the drill bit, and removes drilled out earth. Samples from the mud tell operators what they are drilling through, including oil or gas. The mud is three-quarters water in most cases, the rest being clay, barite, and sometimes special chemicals. The mud is pumped into pits, although in some cases it is simply dumped. The pits can be only a couple of meters (6 feet) deep. In Alaska, reusable mud is recovered and unusable liquids pumped down the well. Drill cuttings, however, are left in the pits, along with scrap solid materials and liquid wastes. Oil production in Prudhoe Bay, which averaged around 1.5 million barrels of oil per day in the 1980s, had by 1990 left behind over 26 million barrels of waste in pits. Bears and wolves moved out of Prudhoe Bay, but overall wildlife populations on the North Slope did not appear to have been diminished.[136]

Directional drilling was adapted in Alaska from offshore operations, where the cost of building platforms for drilling limited the number of wells that could be drilled vertically. Forty wells might be drilled from a single pad using directional drilling. Drilling at up to 60° angles makes possible the drilling of a 2,000 hectare (5,000 acre) field possible from a single pad. However, directional drilling requires an oil-based drilling mud, the disposal of which is an environmental problem.[137] Warm oil flowing up through wells will melt permafrost unless the well casing is insulated. The clustering of wells by means of directional drilling can also cause so much melting that the drilling pad and rig will subside.

A major disaster could occur as a result of sea level rise on the Caspian Sea. The Caspian has risen almost 3 meters (8 feet) between 1978 and 1995, spreading water inland by 15–20 kilometers (9–12 miles). The rising water level will itself create a problem for over 100 villages that could be flooded. But the greatest environmental concern now is that hundreds of pipelines and shut-in wells will be submerged in very salty water. The facilities, not built to withstand corrosion, will create numerous, continuous oil leaks that would present a severe environmental problem.[138]

Cooperation on Caspian fisheries has been minimal since the breakup of the Soviet Union. The sea, which produces four-fifths of the world's sturgeon harvest, is stressed by both oil production and over-fishing. Azerbaijan still allows sturgeon to be taken in the open sea before the fish mature and can reproduce. The official sturgeon catch since the 1991 revolution has fallen 90 percent.[139] Azerbaijan's lack of attention to the Caspian can be attributed to the war for Nagorno-Karabakh. A common fish harvesting policy is sorely needed. Both legal resolution of the status of the body of water for environmental protection and extraordinary cooperation on resource sharing are needed.[140]

Sulfur emissions and disposal associated with oil production may be a concern in the northern Caspian. One analyst predicted that for every barrel of Tengiz oil, 30 pounds of sulfur will be produced, meaning that 10,500 tons of sulfur per day would be produced when production reaches a peak. Washington oil analyst Robert Ebel predicted that Chevron, the principal in Tengiz, would become the "world's largest producer of sulfur."[141]

Pipelines have caused major environmental catastrophes. Gas pipelines cause far less concern for local air and water pollution than do oil pipelines. However, there are exceptions. It was a natural gas liquids pipeline that exploded near a rail line in 1989 and burned to death hundreds of rail passengers. Some unknown percentage of gas leaks from the pipelines, and methane is a powerful greenhouse gas, causing 10–20 times as much radiative forcing per molecule as carbon dioxide.

Petroleum causes significant local water pollution. Washington energy analyst Matthew Sagers has reported on more common spills of several tens of thousands of barrels each, and Transneft, the Russian oil pipeline company, reported in 1999 that Russia spills almost 400,000 barrels of oil per day.[142] A large, more dramatic example occurred in fall 1994 in the Russian republic of Komi. An oil pipeline burst and lost about 730,000 barrels into surrounding ice and snow. The spring thaw allowed the oil to flow toward waterways, and the containment response was slow and poorly organized. The World Bank led an effort to provide an emergency $99 million loan to partially contain the oil.[143]

The Black Sea will be at great risk of a catastrophic oil spill. Oil from the Caspian will initially move by pipeline across the Caucasus to Novorossiysk and to the Georgian port of Supsa where it will be transferred to tankers for shipping through the Bosporus Strait into the Mediterranean. A representative of the Turkish government warned in 1995 that a major oil spill in the Bosporus, which includes "fast currents and twelve sharp turns," and which passes through Istanbul, a city of 12 million, "will stop the flow of all Caspian Sea oil shipments to the west."[144]

Environmental problems plague all refineries, particularly if there is sulfur in the crude oil. Hydrogen sulfide—with the smell of rotten eggs—plus carbon monoxide are serious air pollutants requiring high costs to control. Worse, oily and toxic water discharges into surface waters have already devastated many rivers in the region. The poor health of the Black Sea appears to be at least in part due to refinery discharges along the Dnieper, Danube, and Don Rivers. Sulfur scrubbers and water treatment facilities could be built into refinery upgrades with much less cost as the upgrades are performed rather than waiting to retrofit the refineries later.

Western technology may help. Directional drilling has increased oil companies' ability to minimize the "footprint" of oil wells. A technique called "multiple lateral wells" means that several wells can be drilled from the same spot. That reduces the amount of surface area affected. Closed-loop mud systems reduce the amount of mud waste to be disposed. Western practices may reduce transportation-related spills, and experience from Alaska may be applied to reduce impacts on wildlife and scars on the tundra. Application of such technologies and practices is likely to be voluntary, for better or worse. Russia lacks the tradition of environmental law and the means of citizen enforcement that have been vital to enforcement of western standards. Central Asian republics lack even the protections available in Russia. Any improvement in this potentially catastrophic situation is likely to come only as a result of outside attention and pressure.

Conclusions

Russian historian Nicholas Riasanovsky observed that in democracies, ideas quickly give way to interests. Western involvement in Caspian and Siberian energy geopolitics couples interests and ideas. Symbols of power—state dinners and aid funding—adorn big energy deals in the name of economic reform, energy security, and sustainable economic development. Policy makers equating national security with cheap gasoline run interference for energy companies. Goals of climate and environmental protection, and even of stabilizing Russia's economy, take a back seat. The *Economist* has criticized the predominance of commercial interests in these issues, suggesting that geopolitics has been replaced with "geoeconomics." Harmonizing—making sense of—energy policy priorities requires rethinking interventions in the region's the oil and gas markets for four reasons.

First, governments generally offer financing or other inducements to prevail over other competing nations' industries.[145] The continuing role of government in financing oil development was made clear in June 1998 when the first Russian international financing for a Russian offshore oil drilling project was announced. The U.S. Overseas Private Investment Corporation and the Export-Import Bank of Japan agreed to put up $177 million for a Sakhalin project.[146] An editorial writer observed that all governments now serve as commercial salesmen, suggesting that this is not the best use of leadership.[147] In another editorial about Caspian oil, *The New York Times* said that, "The United States has a nasty habit of subordinating its democratic principles when access to foreign oil reserves seizes the attention of politicians and their corporate benefactors. It happened in Iraq. It should not happen again in the Caspian basin."[148]

Second, military options are very limited in the region, especially in the Caspian Sea. Because the Caspian is landlocked, and because any overland route would go through a war zone or through Russia or Iran, the problem becomes even more geopolitical. That the Caspian is a lake—in reality, if not law—may concern military planners contemplating how to project force into the Caspian in an exercise like Desert Storm, the west's response to Iraq's 1990 invasion of Kuwait. Ongoing debate over the Caspian's legal status as lake or sea will affect rights to build export pipelines under the Caspian even if ownership of the oil itself can be successfully negotiated. That Iran, which the United States considers a rogue state bent on promoting terrorism and developing weapons of mass destruction, straddles the most desirable pipeline routes to the Mediterranean causes conflict both among western allies and between U.S. industry and government. Europe and Russia already work with Iran to develop pipelines. U.S. oil industry lobbyists complain that the United States has put "home country risk" ahead of "U.S. competitiveness abroad," suggesting that the government should step aside and allow private sector cooperation in Iran.[149] But U.S. and Western European energy security is not increased by increasing dependence on a region like the Caspian, which is inherently unstable.

Third, aggressively seeking control of regional oil production is incompatible with international security goals. It may still be possible to promote a "virtuous cycle of investment and profit" in Caspian oil, but current policy exploits conflict for narrow gains and makes oil development a zero-sum game.[150] Industry analysts also write of seeking "non-Russian" opportunities and the advantage of dealing with tyrants instead of Russian democrats.[151] This form of commercial nationalism carries particular significance for the future of Russia. One Clinton Administration official testified before the Senate Foreign Relations committee that U.S. policy in the Caspian was intended to "promote sovereign, prosperous, resilient, democratic states." He mentioned that U.S. policy objectives for the region were "reform . . . , integration into international political and economic institutions, rapid development of Caspian energy resources, with robust U.S. commercial participation . . . cooperative regional relations, and strengthening stakes for Russia and Turkey . . . while isolating Iran."[152] Against this view, Viktor Chernomyrdin, in his first term as prime minister, charged, not entirely for the benefit of the press, that "transnational oil companies are trying to subordinate the Commonwealth of Independent States to their rules of the game, to their economic interests."[153] Aware of these contradictions and concerns, Strobe Talbott cautioned: "In pondering and practicing the geopolitics of oil, let's make sure we are thinking in terms appropriate to the 21st century, not the 19th."[154]

Western officials can justify their involvement in shaping choices that normally would be left to the market—location of wells and pipelines— by invoking national security and even environmental considerations. Care must be taken, however, not to cross an obscure line between shaping markets and creating them. When sector intervention effectively promotes consumption—through subsidies of whatever type—it becomes incompatible with environmental goals. Global climate policy, in particular, is served poorly by the U.S. government using its leverage to provide U.S. consumers with gasoline prices cheaper than water. Much as Soviet apparatchiks have had to rethink the role of government and their own fundamental instincts and beliefs, western champions of energy development may come to realize that the old way of doing business no longer works.

Notes

1. The Caspian Sea area, loosely defined, is the Caucasus and Central Asian regions, including countries ringing the Caspian, plus Georgia and Armenia, which occupy a strategic location between the Caspian and Black Seas. Oil and gas are considered together in one chapter because they sometimes are found together and are exploited by businesses interested in both fuel forms and using similar technologies to develop them. Coal, a third fossil fuel, is considered separately because it is produced by different players using very different technologies. This chapter emphasizes Siberia and the Caspian because little oil is known to exist in the transition economies outside these areas.

2. Charles Clover, "Superpowers Circle Caspian," *Financial Times*, 8 August 1997, p. 4.

3. Bruce Clark, "Oil Fuels American Cheers for an Ex-Communist," *Financial Times*, 3 August 1997, p.3.

4. The horror of 700 impoverished Nigerians burned to death scooping up gasoline from a leaking pipeline is image enough to challenge sanguine visions of energy export as a tool for development. See "Toll Rises to 700 in Nigerian Pipeline Blast," *New York Times*, 21 October 1998, p. 3. See also "Nigeria: A Catastrophe Bound to Happen," *The Economist*, 24 October 1998, p. 44. The writer of the latter story noted, "the fury super-charged by oil. The people of the delta rank among the world's poorest–and oil worth billions of dollars to the government flows through their land," and quoted a local resident as saying, "'The oil companies have been here for 30 years and look at this place.' . . . The village has no school, no clinic, no power and little hope."

5. Frank C. Alexander, Jr., "Caspian Reserves Luring Operators," *Oil & Gas Journal* 95, no. 29 (21 July 1997): 23–30.

6. Jonathan Elkind, Director for Russian, Ukrainian, and Eurasian Affairs, U.S. National Security Council, personal communication, Washington, D.C., 23 October 1998.

7. British Petroleum, *1998 Statistical Review of World Energy*, June 1998 at www.bp.com.

8. Nicholas V. Riasanovsky, *A History of Russia*, 2d ed. (New York: Oxford University Press, 1969), 366–367.

9. Production refers to production, refining, and sale, as opposed to drilling. Titusville, Pennsylvania, holds title as first to drill an oil well.

10. Hugh Pope, "U.S. Report Says Caspian Oil Deposits Might Be Twice as Large as Expected," *Wall Street Journal*, 30 April 1997, p. A10.

11. These data and comparisons, as well as much of the data in this section, were derived from British Petroleum, *1998 Statistical Review of World Energy*, June 1998 at www.bp.com.

12. U.S. Congress, Office of Technology Assessment, *Technology and Soviet Energy Availability* (Washington, D.C.: Government Printing Office, 1981) [hereinafter OTA 1981].

13. Ibid.

14. Matthew J. Sagers, "The Energy Industries of the Former USSR: A Mid-Year Survey," *Post-Soviet Geography* SSSIV, no. 6 (June 1993): 347–355.

15. OTA 1981.

16. Ibid.

17. Ibid., 52; Office of Technology Assessment, *Fueling Reform: Energy Technologies for the Former East Bloc* (Washington, D.C.: Government Printing Office, 1994), 44.

18. These are, in the Russian system, explored reserves that include so-called categories A, B, and C_1; initial estimated reserves, prospective, possible and, possible but less likely reserves are C_2, D_0, D_1, and D_2, respectively.

19. See, for example, James W. Clarke, U.S. Geological Survey, World Energy Resource Program, in *Oil and Gas Resources of the Fergana Basin*, Energy Information Administration, U.S. Department of Energy, DOE/EIA-TR/075, Table 1, p. 3, Washington, D.C., January 1995.

20. International Energy Agency, *Energy Policies of the Russian Federation* (Paris: Organisation for Economic Cooperation and Development, 1995).

21. British Petroleum, *Annual Statistical Review, 1999,* at www.bp.com.

22. International Energy Agency, *Energy Policies of Ukraine* (Paris: Organisation for Economic Cooperation and Development, 1996), 119.

23. Ibid.; U.S. Embassy Kyiv, *Ukraine's Oil and Gas Outlook–Difficult Horizons,* 31 October 1996, http://www.iep.doc.gov/bisnis/cables/961101up.htm. Embassy staff also quote Bogdan Babiev, Chairman of the Ukrainian State Committee on Oil and Gas, at the Crimea–96 Conference, Yalta, 7–9 October 1996.

24. IEA, *Energy Policies of Ukraine,* 147.

25. IEA, *Energy Policies of the Russian Federation,* 106.

26. E. M. Khartukov, *Russia's Oil Prices: Passage to the Market,* East-West Center Working Papers, Energy and Minerals Series, no. 20. (September 1995), 1–8; Mark A. Gyetvay, "Restructuring, Consolidation Top Solutions for Russia's Major Oil Companies' Woes," *Oil & Gas Journal* (31 March 2000): 20–26.

27. Eugene M. Khartukov (International Center for Petroleum Business Studies), "Low Oil Prices, Economic Woes Threaten Russian Oil Exports," *Oil & Gas Journal,* 8 June 1998, 25–30.

28. Pope, "U.S. Report Says Caspian Oil Deposits Might Be Twice as Large as Expected," A10.

29. Charles Clover and Robert Corzine, "Treasure Under the Sea," *Financial Times,* 1 May 1997, p. 11.

30. Ibid., quoting "a western oil executive"; Robert Corzine, "Cash Is Gone, Please Find a Lot of Oil," *Financial Times,* 10 November 1999, p. 13.

31. Dan Morgan and David B. Ottaway, "Vast Kazakh Field Stirs U.S.-Russian Rivalry; Pipelines Are Key to American Exports," *Washington Post,* 6 October 1998, p. A01.

32. Robert E. Ebel, *Energy Choices in the Near Abroad: The Haves and Have-nots Face the Future* (Washington, D.C.: Center for Strategic and International Studies, 1997), 19.

33. Ebel, 90 and 94.

34. U.S. Department of Energy, "Kazakhstan," Country Analysis Briefs, http://www.eia.doe.gov/emeu/cabs/kazak.html, April 2000.

35. Michael R. Gordon, "Pact to Build Caspian Oil Pipeline Reached by Russia, Kazakhstan, Companies," *New York Times*, 25 November 1998.

36. Pennzoil in 1999 indicated plans to sell its interest in this contract.

37. Clover and Corzine, "Treasure Under the Sea," 11.

38. Ebel, *Energy Choices in the Near Abroad*, 122–123. For a general discussion of pipelines and the Caspian, see K. M. Kerimov, Ch. M. Khalifazade, F. M. Hajiyev, I. S. Hasanov, and M. M. Zeynalov, "Construction of the Main Pipelines and Their Operation with Considered Geodynamic and Engineering Geology Aspects Within the Caspian Basin–Black Sea Region" (paper presented at the Fifth International Congress on Energy, Economy, and Ecology, Baku, 24–26 September 1999).

39. Ebel, *Energy Choices in the Near Abroad*, 121.

40. United Nations Population Division, *World Population Prospects 1950–2050*, (New York: United Nations, 1996).

41. Robert Corzine, "Low Prices Hit Big Kazakh Oil Project," *Financial Times*, 20 May 1998, p. 7.

42. Einar Tresselt, Vice President, Azerbaijan International Operating Company, personal communication, Baku, Azerbaijan, 8 October 1998. Tresselt recalled his experience with an offshore oil platform in the North Sea that cost $1 billion, roughly the same as the AIOC investment, and produced 75,000 barrels per day, the current production rate for AIOC in the Caspian. If one assumes a 10 percent discount rate, the capital investment would amount roughly to $5 per barrel of oil.

43. Charles Clover and Carlotta Gall, "Kazakhstan Free to Drill Huge Caspian Oilfield After Signing Pact with Russia," *Financial Times*, 7 July 1998, p.4.

44. The James A. Baker III Institute of Public Policy at Rice University forecast that Caspian oil production would by 2010 reach no more than 3.5 million barrels per day. The institute cautioned that the demand and output may justify only one export pipeline. James A. Baker III Institute of Public Policy, Rice University, *Unlocking the Assets: Energy and the Future of Central Asia and the Caucasus. A Political, Economic, and Cultural Analysis*, Rice University (April 1998), at www.riceinfo.rice.edu/projects/baker/publications.

45. Sagers, "Energy Industries of the Former USSR," 372.

46. IEA, *Russian Energy Policies*, 141.

47. Mikhail G. Rudin and Valeri S. Plotnikov, "Russia Struggles Toward Refinery Modernization Goals," *Oil & Gas Journal*, 18 August 1997, pp. 31–35.

48. British Petroluem, *Annual Statistical Review 1999*.

49. Poet Andre Kodrescu derides the notion of the changes as "revolutionary." But if revolution means drastic change, this one qualifies.

50. Sagers, "Energy Industries of the Former USSR," 374.

51. Former Deputy Energy Secretary William White, personal communication, Washington, D.C., October 1995.

52. OTA, *Fueling Reform*, 46.

53. IEA, *Energy Policies of Ukraine*, 129.

54. Ibid., 133.

55. British Petroleum, *Annual Statistical Review, 1998*.

56. John Thornhill, "Gazprom Hit by Tax Penalties," *Financial Times*, 11 July 1997, p. 23.

57. Chrystia Freeland, "Gazprom Plans to Save $522M in Shake-up," *Financial Times*, 16 April 1997, p. 1.

58. Gazprom produced over 21 exajoules in 1995, compared to a combined gas and oil total of 6.1 exajoules for Exxon. International Energy Agency, *Energy Policies of the Russian Federation*, 49; Goscomstat, *Налияие Легковых Автомоьилеи вСобствинности Граждан по Гегиогам Российской Федерации* (Moscow: Russian Federation, 1997).

59. OTA, *Fueling Reform*, 27.

60. OTA 1981, 57.

61. Ebel, *Energy Choices in the Near Abroad*, 114 and p. 95.

62. Alexander, "Caspian Reserves Luring Operators," 23–30.

63. Russian gas pipeline export access to Europe for Turkmen gas was cut to 8.2 billion cubic meters in 1994. Ebel, *Energy Choices in the Near Abroad*, 128; "Turkmenistan, Ukraine Try to Resolve Gas Non-Payment Problems," *RFE/RL Newsline* 3, no. 78, pt. I (22 April 1999); http://www.rferl.org/newsline/1999/04/220499.html. Ukraine in 1999 owed Russia over $1.6 billion for natural gas, and Russia accused Ukraine of stealing 2.5 billion cubic meters of gas (0.1 exajoules) intended for transit through Ukraine to Western Europe. Agence France Presse (wire service), "Stop Stealing Our Gas, Gazprom Chief Tells Ukraine," 12 January 1999. The International Monetary Fund accused the national gas utility, Neftehaz Ukraini, of using barter deals to sell gas to avoid taxes. The utility in 1999 owed almost $1 billion to the Ukrainian government.

64. Ebel, *Energy Choices in the Near Abroad*, 131.

65. Energy Information Administration, U.S. Department of Energy, "Russia: February 2000," http://www.eia.doe.gov/emeu/cabs/russia.html; IEA, *Energy Policies of the Russian Federation*, 158.

66. IEA, *Energy Policies of Ukraine*, 129.

67. Halia Plachynda, "Odessa Connection: An Oil Terminal under Construction Near Odessa May Be the Key to Free Ukraine from Russian Oil, But the Billion-dollar Facility Has Only $23 Million in Construction Funds," *Intelnews Business Journal*, 18 August 1997.

68. Anonymous, "Chechnya Threatens to Halt Oil Transit," *RFE/RL Newsline* 2, no. 46, pt. I (9 March 1998); http://www.rferl.org/newsline/1998/03/090398.html. See also Yo'av Karny, "The Chechen Leaders Thrive On Perpetual, Idealized War," *Washington Post*, 10 October 1999, p. B1.

69. Anthony Robinson and Selina Williams, "Falling Price of Oil Spotlights Perils of Caspian Investments in the Pipeline," *Financial Times*, 27 July 1998, p. 3.

70. Carlotta Gall, "Nagorny Karabakh: Peace Deal Eludes Ethnic Conflict," *Financial Times*, 3 March 1998, p. 14; Steve LeVine, "Instability by the Barrelful: Central Asia's Coming Oil Bonanza and Its Consequences," *New York Times*, 17 February 1998, p. D-1; Stephen Kinzer, "In Post-Soviet Era, the Old Style Works: Azerbaijan Autocrat Quiets the Chaos, But Democracy Suffers," *New York Times*, 21 September 1997, p. 16.

71. Ebel, *Energy Choices in the Near Abroad*, 112.

72. Robert Corzine, "Extra Capacity for Caspian Sea Pipeline Being Considered," *Financial Times*, 21 February 1997, p. 7.

73. Frank C. Alexander, Jr., "Caspian Reserves Luring Operators," *Oil & Gas Journal*, 21 July 1997, pp. 23–30.

74. Ebel, *Energy Choices in the Near Abroad*, 73 and 80.

75. Ebel, *Energy Choices in the Near Abroad*, 133–134.

76. *RFE/RL Newsline* 2, no. 161, pt. I (21 August 1998); http://www.rferl. org/newsline/1998/08/210898.html.

77. Alexander, "Caspian Reserves Luring Operators," 28.

78. "U.S.-Uzbekistan Joint Commission: Trade/Investment/Energy," State Department press release, 27 February 1998.

79. Alexander, "Caspian Reserves Luring Operators," 28.

80. Ebel, *Energy Choices in the Near Abroad*, 165–166.

81. Paul Betts and Robert Corzine, "Eni and Gazprom to Build $3bn Pipeline," *Financial Times*, 4 February 1999, p. 1.

82. "Chinese Company Wins Kazakh Tender," *RFE/RL Newsline* 1, no. 46, pt. I (5 June 1997); http://www.rferl.org/newsline/1997/06/050697.html.

83. Gary Dirks, President, BP China, personal communication, Beijing, China, 18 November 1997. For resource estimates in eastern Russia, see R. N. Diyashev, E. S. Ziganshin, and V. N. Ryabchenko, "Verkhnechonsky Field Shows Eastern Russia's Potential," *Oil & Gas Journal*, 19 July 1999, p. 53.

84. Tony Walker, "Sino-Russian Fuel Deal Close," *Financial Times*, 19 June 1997, p. 1.

85. Agence France Presse (wire service), "BP Pulls Out Top Executive from Russia's Sidanko," 16 November 1999.

86. Charles Clover and Robert Corzine, "Caspian Oil Share Sought by Russians," *Financial Times*, 14 July 1997, p. 4.

87. Irvin Molotsky, "U.S. Is Said to End Opposition to Iran Pipeline," *New York Times*, 27 July 1997, p. 3.

88. "Crackdown in Iran," *New York Times*, 11 October 1999, A22; Aliyev Rafig Yahya oglu, founder and director of "Irshad" Center for Islamic Studies, personal communication, Baku, 21 September 1999.

89. George Rogers Jackson, "The Politics of Oil Investment: Regime-Type and Foreign Direct Investment in the Russian and Kazakh Oil Industries" (Faculty of the Graduate School of the University of Texas at Austin in Partial Fulfillment of the Requirements for the Degree of Master of Arts, The University of Texas at Austin May, 1997).

90. Ahmed Rashid, "As Washington Says It Won't Oppose a Gas Pipeline Across Iran and China Invests in Kazakhstan's Oil Industry, the Door to Central Asia's Energy Resources Is Opening," *Far Eastern Economic Review* 160 (14 August 1997): 42–44.

91. Robert Corzine, "Iran Celebrates $200m Caspian Gas Pipeline," *Financial Times*, 30 December 1997, p. 1.

92. Statement by the Press Secretary, "President Clinton Names Ambassador Richard L. Morningstar to be Special Advisor to the President and the Secretary of State for Caspian Basin Energy Diplomacy," The White House, 24 July 1998.

93. The transport cost will total about $0.80 per GJ, compared to an export price of about $3 per GJ. Gazprom has borrowed $2.5 billion from the German

Dresdner Bank to finance construction. *OMRI Daily Digest* I, no. 18 (27 January 97); www.friends-partners.org/friends/news/omri/1997/01/970127.html.

94. Roula Khalaf and Robert Corzine, "Tehran To Issue $400m Caspian Pipeline Tender," *Financial Times*, 29 May 1998, p. 6.

95. Bruce Clark, "Putting Down Routes: Ebb and Flow in Caspian Pipeline Politics," *Financial Times*, 26 March 1998, p. 6.

96. Kelly Couturier, "US Eximbank Willing to Finance Caspian Pipelines," *Financial Times*, 29 May 1998, p. 6.

97. William H. Courtney (Special Assistant to the President and Senior Director for Russia/Ukraine/Eurasia, National Security Council), "U.S. Policy on Caspian Energy" (paper presented at the Central Asian Institute, School of Advanced International Studies, Johns Hopkins University, Washington, D.C., 4 February 1998).

98. U.S. Department of Energy, "Oil Companies," at www.itaiep.doc.gov/bisnis/isa.

99. Natalia Gurushina, "Oil Joint Ventures in Russia in 1996," *RFE/RL Newsline* (3 April 1997); http://www.rferl.org/newsline/1997/03/1-RUS/rus-280397.html.

100. Martha Hamilton, "Russian Firm Opens a Va. Gas Station," *Washington Post*, 29 July 1997, p. C2.

101. Minatep merged with Most and Uneximbank banks during the August 1998 financial crisis.

102. Sagers, "Energy Industries of the Former USSR," 346.

103. Michael R. Gordon, "Oil-Price Drop Threatens to Squelch Russia's Economic Rebound," *The New York Times*, 20 March 1998, p. A5.

104. OTA, *Fueling Reform*, 43.

105. Dale F. Gray, *Evaluation of Taxes and Revenues from the Energy Sector in the Baltics, Russia, and Other Former Soviet Union Countries*, Working Paper 98/34 (Washington, D.C.: International Monetary Fund, 1998).

106. International Monetary Fund, "Progress with Fiscal Reform in Countries in Transition," in *From Stabilization to Sustainability* (Washington, D.C.: International Monetary Fund, 1998), Chapter V.

107. Michele Bertelli, "Investment in Poland's Refining Industry: A Long-Term Opportunity for Conoco" (in partial fulfillment of Masters degree requirements, Johns Hopkins University School of Advanced International Study, Washington, D.C., May 1997), 1–3.

108. See, for example, "Poland, Norway Agree to Build Gas Pipeline," *RFE/RL Newsline* 3, no. 88, pt. II (6 May 1999); http://www.rferl.org/newsline/1999/05/060599.html.

109. Bertelli, "Investment in Poland's Refining Industry," 9, citing *Enerfinance*.

110. Bretislav Karásek and Miroslav Cibulka, "Czech Crude Oil Supplies Diversified by New Pipeline," *Oil & Gas Journal*, 25 August 1997, pp. 84–88.

111. Reuters, "$10bn Gazprom Deal Finalised," *Financial Times*, 19 October 1998, p. 5.

112. Cambridge Energy Research Associates, "Gazproms' Future: Structure & Strategy," prepared for the U.S. Department of Energy, June 1995. See also *Oil and Gas Journal*, June 12, 1995, pp. 27–32.

113. Associated Press News Service (wire service), "Russia's Gazprom Production Up," 27 April 1999.

114. See generally Jonathan P. Stern, *The Russian Natural Gas "Bubble,": Consequences for European Gas Markets* (London: Royal Institute of International Affairs, 1995).

115. Cambridge Energy Research Associates, "Gazprom's Future: Structure & Strategy."

116. Frederick Stüdemann, "Gazprom Extends Reach in Germany As Liberalization Looms," *Financial Times*, 2 July 1998, p. 3.

117. Anonymous, "Gazprom Turns Down the Gas on Ukraine," *Financial Times*, 23 July 1997, p. 3.

118. International Energy Agency, *Energy Policies of Ukraine: 1996 Survey* (Paris: Organization for Economic Cooperation and Development, 1996), 117.

119. Ibid.

120. Ibid., 124.

121. Ibid., 125.

122. Viktor Luhovyk, "Ukrgazprom Calling for More Changes in JKX Top Management," *Kyiv Post*, 2 February 1998, http://www.thepost.kiev.ua/archive/ar-07is011-10feb1998.txt.

123. Eva B. Pregon, "What Does Gazprom's Restructuring Mean for Investors?" (in partial fulfillment of Masters degree requirements, Johns Hopkins University School of Advanced International Study, Washington, D.C., May 1997) 2, citing Chrystia Freeland, "Adjustments to Bear," *Financial Times*, 28 April 1997; Pregon, 3, citing John Thornhill, "Economic Reform Crucial," *Financial Times*, 9 April 1997; John Thornhill, "Gazprom Hit by Tax Penalties," *Financial Times*, 11 July 1997, p. 23; CENEf, *Russian Energy Picture* (Moscow: Center for Energy Efficiency, October–December 1997), 6.

124. Thornhill, "Gazprom Hit by Tax Penalties," 23.

125. Chrystia Freeland, "Gazprom Plans to Save $522m in Shake-up," *Financial Times*, 16 April 1997, p. 1.

126. Pregon, "What Does Gazprom's Restructuring Mean?" 3, citing Chrystia Freeland, "Russian Leadership Warns of a Budget Crisis," *Financial Times*, 18 April 1997.

127. Ibid., 5, citing *Izvestia* interview, cited in *The Current Digest of the Soviet Press*, 16 April 1997. Ibid., citing John Thornhill, "Yeltsin Moves to Overhaul Russia's State Monopolies," *Financial Times*, 29 April 1997.

128. Ibid., 10; Gazprom Chairman Rem Vyakhirev, quoted in "Big Gas Firm Proves a Liability for Yeltsin," *Wall Street Journal*, 5 March 1996, sect. A, p. 11.

129. Pregon, "What Does Gazprom's Restructuring Mean?" 7, citing *Soviet Press Digest*, 17 April 1997.

130. *Russian Energy Picture*, Center for Energy Efficiency, Moscow, January–March 1997, p. 19.

131. "Decree Lowers Gas Rates for Some Domestic Customers," *RFE/RL Newsline* 1, no. 57, pt. I (20 June 1997); http://www.friends-partners.org/friends/news/omri/1997/06/970620I.html.

132. "Gazprom: Pipe Dreams," *The Economist*, 5 July 1997, p. 63. Gazprom has demonstrated that it can raise capital in the domestic Russian market by selling

bonds, even after the disastrous financial collapse of 1998. Reuters (wire service), "Gazprom Issues First Russian Corporate Bond," 13 July 1999.

133. Many of the issues touched on here are treated in some detail in Chapter 7.

134. William Eichbaum, Vice President, World Wildlife Fund, personal communication, Washington, D.C., 7 March 2000; Vladimir Krever, Olga Pereladova, Margaret Wiliams, and Harmut Jungius, *Biodiversity in Central Asia: An Analysis of Biodiversity and Current Threats and Initial Investment Portfolio* (Moscow: Text Publishers, 1998); and *Circumpolar Protected Area Network*, WWF Arctic Bulletin no. 2.95, 6–13.

135. U.S. Congress, Office of Technology Assessment, *Oil Production in the Arctic National Wildlife Refuge: The Technology and the Alaskan Oil Context*, OTA-E-394 (Washington, D.C.: Government Printing Office, February 1989), 40.

136. Ibid., 58.

137. Ibid., 43–44.

138. Ebel, *Energy Choices in the Near Abroad*, 100.

139. Russell Dyk, personal communication, Johns Hopkins University, 4 March 1997, citing Vadim Birstein, "Sturgeon May Soon Disappear from the Caspian Sea," *Russian Conservation News* no. 7 (Spring 1996).

140. Ibid.

141. Ebel, *Energy Choices in the Near Abroad*, 114–115.

142. Sagers, "Energy Industries of the Former USSR," 377.

143. "World Bank Approves Loan to Clean up Komi Oil Spill," *OMRI Daily Digest* I no. 71 (10 April 1995); http://www.friendspartners.org/friends/news/omri/1995/04/950410I.html.

144. Reported by Ebel, *Energy Choices in the Near Abroad*, 53.

145. Unsigned editorial, "Don't Be Salesmen," *The Economist*, 1 February 1997, p. 17.

146. The project would involve the "Ms": Marathon, Mitsui, and Mitsubishi, as well as Royal Dutch/Shell. It would produce 45,000 barrels of oil per day, six months per year. See Robert Corzine, "Russia Sets up Funding for Oil Project," *Financial Times*, 12 June 1998, p. 34.

147. An example of the new geoeconomics was the U.S. hosting of Azerbaijani President Heydar Aliyev. The dinner highlighted a curious confusion in U.S. policy. Driven by a strong Armenian-American lobby, U.S. law forbids humanitarian assistance to Azerbaijan, making that country one of the very few in the world achieving such ignominious status. See Bruce Clark, "Oil Fuels American Cheers for an Ex-Communist," *Financial Times*, 3 August 1997, p. 3.

148. Anonymous, "Caspian Oil and Quicksand," *The New York Times*, 4 August 1997, p. A18.

149. John H. Lichtblau (Chairman, Petroleum Industry Research Foundation, Inc.), "U.S. Caspian Area Foreign Policy in Conflict with Resource Plans," *Oil & Gas Journal*, 11 August 1997, pp. 19–22.

150. Anonymous, "Soft Power," *Financial Times*, 14 August 1997, p. 11.

151. See, for example, Elshan Alekberov, "Despite Political Obstacles, Energy Work Progresses Around Caspian Sea," *Oil & Gas Journal*, 15 June 1998, pp. 38–47. Alekberov introduces his article by writing about the frustration of dealing with Russia: "Virtually all international oil giants and a number of independents, fol-

lowed by service companies, are in or are fiercely trying to get projects in the non-Russian Caspian states of the former Soviet Union. The countries lag well behind Russia in the development of democratic institutions and irreversible guarantees for individual freedoms. But their strong central governments and autocratic leaders have provided both legal frameworks and investment protection sufficient to permit major oil and gas projects to go forward with foreign partners" (p. 38).

152. Testimony of Stuart Eizenstat, Under Secretary of State for Economic and Business Affairs, U.S. Department of State, before the Senate Foreign Relations Committee, 22 July 1997.

153. Unsigned article, "Premier Blasts Western Oil Policies," *Moscow News*, 14 May 1996, p. V.

154. Deputy Secretary of State Strobe Talbott, quoted in unsigned editorial, "Caspian Oil and Quicksand," *The New York Times*, 4 August 1997, p. A18.

5

Coal and Reform

Eastern Europe lacks oil and gas resources but has in abundance what Chilean poet Pablo Neruda called "the disaster of coal."[1] Coal is difficult to mine, transport, and burn. In the absence of both market discipline and public accountability, and without the technology developed elsewhere in response to such pressure, coal use in the region before the end of central planning caused devastating environmental, safety, and social problems. Coal-fired power plants across the region still cause millions of cases of respiratory disease annually, even stunting children's growth. Coal mining accidents annually cause more death in Ukraine—and in Russia—than Chernobyl was responsible for in a decade. Regional coal-sector job losses total in the hundreds of thousands and will run higher before restructuring is complete.

Regional mine worker productivity is notoriously low. Ukrainian miners produce 15–60 times less coal per worker than U.S. miners. Coal mining safety has been catastrophically neglected in the region. Approximately 300 people die in coal mining accidents in Russia each year; almost 300 more die in Ukraine.[2] The death rate for coal miners in Ukraine is four per million tons of coal produced, 100 times higher than in the United States.

Coal is quite literally dirty, being 15–50 percent ash, and contains compounds of silicon, aluminum, and iron—the constituents of dirt—along with varying amounts of sulfur, arsenic, lead, and antimony. Environmental pollution caused by coal is extreme. Journalists visiting Eastern Europe once wrote of "darkness at noon," a mixture of cloud and dust and soot so thick that headlights were required for daytime driving (see Figure 5.1). It was coal pollution they were describing. In Krakow in the late 1980s, white clothing worn outdoors would be black in hours. In the Czech Republic, brown coal was composed half of dirt and half of combustible carbon material. The particulate matter emitted from brown coal

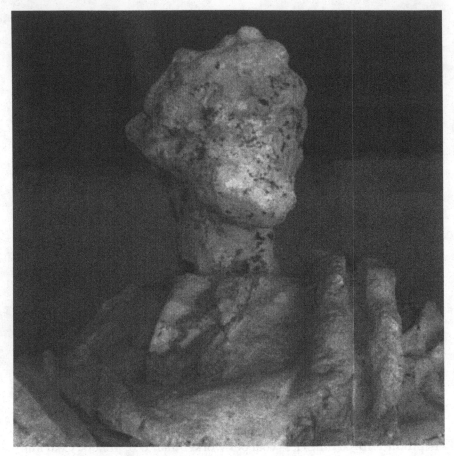

Figure 5.1 The Effect of Acid Rain
Photo by Ecofund Poland.

burning utilities contained heavy metals in such concentrations that they
stunted bone growth in children by six months.[3] When burned in household stoves or industrial furnaces with low stacks, pollutants quickly set-
tle on the population and countryside. When burned in power plants
with taller stacks, sulfur dioxide produced in combustion is lofted higher
into the atmosphere, where it mixes with water to produce acid rain. In
the late 1980s, rain in the "Black Triangle" region where Poland, the
Czech Republic, and Germany meet had a pH of under 4 in the 1980s, or
the acidity of tomato juice. In Romania, coal was of such poor quality—

contained so much non-carbon material—that power plants could not burn it.

Coal sector reform has also been slow in western nations, including Germany, where subsidies for mining were sustained until the late 1990s; Great Britain, where natural gas use in new power facilities was suspended as a way of subsidizing coal; and the United States, where recently the state-owned Tennessee Valley Authority, along with several private utilities, were charged with the largest air pollution violations in thirty years, due to their use of coal. The pain of discharging workers and the cost of cleaning up coal's pollution, plus considerations of energy security associated with importing substitutes for coal, has made for slow reform. Reform has been the subject of considerable attention of domestic governments and international agencies, particularly the World Bank, but cleaner air and higher productivity have come about mainly as the result of depressed demand.

Coal's Decline

Coal is the world's most abundant fossil fuel, and Russia's coal reserves of 265 billion tons alone are enough to supply the entire region for 440 years. Added to that vast total are resources totaling about 50 billion tons each in Kazakhstan, Poland, and Ukraine. The countries most dependent on coal are the Czech Republic, Poland, Kazakhstan, and Ukraine. Kazakhstan is a special case, being an oil- and gas-rich nation.

Russia, Poland, Ukraine, Kazakhstan, and the Czech Republic together produce about 573 million tons of coal each year, accounting for 95 percent of regional production (see Table 5.1). The average market share for coal around the world is about one-quarter of total energy supply. Coal is heavily used for making electricity, generating 98 percent of Polish and 64 percent of Czech power. These two nations, however, plan to reduce coal use to reduce environmental pollution.

Coal use will continue dropping as consumers move to higher-grade fuels or as better coal seams are depleted. Total coal production in the region is down 42 percent since 1988, far more than the drop in oil production (see Figure 5.2). Poland's coal use has already fallen by 28 percent over the last decade, a total of 62 million tons. Coal use in that nation has declined faster than the use of imported oil and gas, as coal's share of primary energy supply fell from 80 to 70 percent in a decade. That share, if not total coal consumption, is likely to fall further. Russian coal production and use resembles the collapse of oil in that country, having fallen from 336 million tons in 1990 to 210 million tons in 1997.[4] Coal output is down in Romania by 50 percent, in the Czech Republic by 40 percent, and in Ukraine by 60 percent.

TABLE 5.1   Coal Production in the Transition Economies, 1999

Country	Million Tons	Regional Share (percent)
Russian Federation	197	35
Poland	128	26
Ukraine	74	13
Kazakhstan	52	12
Czech Republic	37	9
Romania	9	2
Bulgaria	8	2
Hungary	7	1
Other Regional	2	<1
Regional Total	560	100

NOTE: Coal assumed to average 24 gigajoules per ton.

SOURCES: British Petroleum, *Statistical Review of World Energy, 2000*; BPAmoco, "Statistical Review of World Energy," http://www.bpamoco/worldenergy/.

Ukraine's coal industry in the Donets Basin, or Donbas, has a history of over 125 years. Production peaked in the 1970s at just under 200 million tons annually, when Soviet planners began shifting investment to the Kuznets Basin, or Kuzbas, in Siberia, where thick seams of coal could be surface mined. The economic crisis of the 1990s accelerated the Ukrainian coal industry's decline, which was at least in part due to rising production costs and falling coal quality. Ukrainian coal use has fallen by half since the late 1980s to about 65 million tons, roughly the same level of consumption as in Poland. The dire condition of the Ukrainian coal industry is reflected in the fact that Ukraine, which has struggled to close the Chernobyl nuclear power plant and pay for imported natural gas, has cut coal consumption deeply.

There is irony in Kazakhstani coal dependence. Kazakhstan possesses some 50 billion barrels of oil reserves and an equivalent gas resource, yet obtains over half its energy from coal. Kazakhstan uses about 50 percent more coal per capita than the United States.[5] One power plant in Kazakhstan emits almost four times the amount of particulate matter released by all power plants in the United States.[6] Nine of every ten Kazakhstani power stations is coal-fired.

Coal supplies much of the region's extensive district heating and household energy supply, almost exclusively so in Eastern Europe. Be-

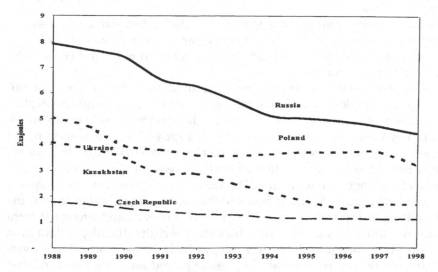

Figure 5.2 Coal Production, Leading Transition Economies, 1988–1999
SOURCES: British Petroleum, *Statistical Review of World Energy, 2000*; BPAmoco, "Statistical Review of World Energy," http://www.bpamoco/worldenergy/.

cause they lack basic controls and emit their pollution in densely populated urban centers, these systems may create health problems out of proportion to their total energy use.

Mining Technology

Coal is mined in the region both from the surface and deep underground. About half of all Russian coal is produced underground, a share that has declined 13 percent since 1980. Most Ukrainian coal is mined underground, most of it bituminous, medium-quality coal. Surface mines account for 70 percent and 75 percent of Kazakh and Czech coal production, respectively, but only one-third of Hungary and Poland's.

Surface mines typically involve thick deposits of brown or relatively low-quality coal, used mainly for power generation. The Bełchatow mine in western Poland is typical. First, the surface was scraped off and the coal seam exposed. Earth movers as tall as skyscrapers were erected, holding rotary buckets that excavated coal and placed it on moving conveyors. The conveyors took the coal directly to one of several "mine-mouth" plants located nearby to burn the coal, produce electricity, and send the power off to the power grid on large transmission lines. This practice is similar to West German technology, but very different from the surface mining common to Appalachia in the United States, where seams are thin—often less than two meters—and where power shovels and even augurs are used to extract coal. Reclamation in Poland is also inferior to German practice.

Coal surface mines are far safer than underground mines because underground explosions and roof falls account for most coal mining deaths, and these obviously are not an issue on the surface. Surface mines are often, however, more environmentally disruptive because everything on the surface—fields, trees, homes—must be removed. German technology has permitted the reclamation of surface areas for agricultural use and even for village reconstruction. The technology is expensive and requires considerable care. A critical factor in the success of soil reclamation is the handling of top soil. It must be carefully removed and protected from compaction and leaching, which destroy fertility. Ideally, reclamation takes place concurrently with mining, which means that soil and overburden removed from an active site are placed on a site where mining has just been completed. In Eastern Europe and Russia, this practice has not been followed. Bełchatow is more typical, where a mountain of infertile spoil has accumulated next to the large open pit. This practice not only removes the possibility of returning the land to use after mining is completed but creates an eyesore and a source of erosion and leachate, and can even cause small tremors.[7] Underground mining in Eastern Europe and the former Soviet Union also differs substantially from U.S. practice, although it is more consistent with European technology. The "long wall" mining technique accounts for 85 percent of the region's underground coal production.

In U.S. deep mines, excavation begins with a vertical or slanted shaft down to the level of the coal. Rectangular room-shaped sections are then mined by leaving pillars of coal to support the roof: hence the name, room and pillar mining. Coal is extracted with special drills, hammers, and loaders, and the coal is carried to the surface in special hopper cars. Mine ceilings between the pillars are supported with roof bolts and timbers.

Long wall mining is started also by sinking a shaft, but instead of mining rooms supported with pillars, which can leave behind as much as 40 percent of the coal, a tunnel is cut in a rectangular shape, surrounded by an area of coal on all sides. Then, starting at one corner of the rectangle, a rotary saw-like device is mounted on the face of the coal. The saw then passes back and forth along one edge of the rectangle, cutting coal off as it goes. As the drill moves forward along the coal seam face, coal falls off onto a conveyor and is carried to the surface. The roof behind the drill, or saw, is supported by movable hydraulic jacks. These are moved forward as the mining progresses, allowing the roof to collapse behind them. When the drill reaches the opposite side of the rectangular section, the drill is moved forward to where another section has been prepared.

This approach should be much safer than room and pillar mining. First of all, the chance of a roof fall—the local collapse of rock—should be much reduced because the roof is supported by hydraulic jacks. Second, the controlled collapse of the roof stabilizes the mine. This has benefits on the surface above the mine. With room and pillar mining, surface subsidence is more severe because the surface subsides unevenly when the roof collapses between pillars. Because long wall mines do not leave pillars, surface damage is more controlled and less extreme. Long wall technology is inherently more prone to automation, and should reduce the number of workers required at the dangerous mining face, where accidents occur involving the heavy moving equipment.

Unfortunately, theory is far from practice. Ukraine's industry provides the most egregious example. Equipment at the working faces runs much less than full time. The equipment itself is outdated, and conveyor belts, fans, pumps, electrical equipment, and ventilation systems are not maintained.[8] These problems stem partly from declining quality of Ukrainian coal reserves. The mines must generally be dug 1,000 meters (3,300 feet) below the surface. The seams tend to be inclined, which makes them harder to follow and extract. The mines are gassy and contain troublesome amounts of saltwater, which when dumped causes severe water pollution.

The worst safety problem in Ukrainian and Russian deep mines is methane gas. The coal seams are unusually gassy. Safety can be enhanced and useful energy sometimes produced by first drilling the coal seams to drain the methane, but usually it is simply vented from the mine shaft into the atmosphere. Current practice has led to frequent explosions, and 250–300 miners are killed each year in Russia and Ukraine as a result. This high rate of death has continued since 1988 in Russia despite coal production having fallen by 40 percent, meaning that the safety of Russian coal mining has seriously deteriorated from an already poor position.[9] The number of dead in Russia's coal mines increased from 172 in 1996 to 277 in 1997.[10]

Coal Pollution

Coal mining and burning causes a plethora of serious environmental problems. Land impacts from both deep and surface mining can be severe. Air pollution from burning coal is probably the most serious energy-related public health threat in the region. Air pollution in the form of acid rain also causes serious economic damage to crops, buildings, and equipment. In addition, coal produces more carbon dioxide, a serious greenhouse gas, per unit of useful energy than any other fuel.

Coal-related pollution in the late 1980s was particularly severe in Poland, where it was concentrated in Upper Silesia and in the Krakow region. Power plants, district heating, and small furnaces, boilers, and stoves all contributed, although two-thirds of the sulfur dioxide originated in eight coal-fired power plants. The International Energy Agency pointed out that in 1993, the Bełchatow lignite-fired power complex of 4,000 megawatts in western Poland produced more sulfur dioxide than all of Sweden, Norway, and Denmark combined.[11]

The World Bank has estimated that sulfur and nitrogen oxide emissions in the region cause health and property damage that totals $200–700 per ton of pollutant.[12] Given 1990 emissions levels in Poland, the coal-related pollution cost was in the range of $2–7 billion per year. The health and materials costs thus averaged $50–200 per person per year. Fortunately, reform has meant reduction in emissions. Sulfur dioxide and particulate emissions are down 40 and 55 percent, respectively.[13] The largest particulate problems were in Katowice and Krakow, and these cities saw the most dramatic improvements.[14] Some of this improvement was due to improved fuel quality, which for hard coal reduced the ash content from 28 to 22 percent in the five-year period after 1988, whereas sulfur content dropped from 1.15 to 0.83 percent. Small improvements were realized in soft coal, or lignite (see Figure 5.3 and Box 5.1). Consumption of soft coal fell little over the period, but hard coal use fell by one-third. To some extent, improvement in particulate emissions was made by improving maintenance and performance of particulate removal systems—electrostatic precipitators, primarily—but economic decline and reform appear to have accounted for most of the change.

Emissions in Ukraine for all the main categories—particulates, sulfur dioxide, nitrogen dioxide, and carbon dioxide—were down from 1990 levels by about 40 percent by 1994, reflecting not an improvement in environmental management but rather the economic decline.

The Czech Republic recorded particulate emission reductions of 70 percent between 1990 and 1995 (see Box 5.2). This result—given the stable level of economic output in that country—represented one of the largest environmental payoffs in the region. Sulfur emissions were also down by 43 percent over the same period. These improvements were at least in part the result of a conscious policy of the Czech government. Bulgarian emissions for the four main categories were down 20–30 percent, more consistent with the economic depression. Russian emission re-

---

BOX 5.1   REDUCING COAL CONSUMPTION IN POLAND

The Bush Administration launched a major effort to demonstrate U.S. clean coal technology at the Skawina Power Station in Krakow, Poland. The project was initially slated as a $20 million effort to reduce pollution levels in Krakow, which is a city of significant historical, cultural, and tourist value. Krakow's burden of particulate emissions and acid rain have caused great harm to both the historic artifacts of the city and the health of its population.

The Skawina project in its early stages was interpreted as a "scrubber retrofit project," which Polish environmentalists criticized because they felt it did more to promote U.S. business than to address pollution in the city. It was estimated that the $20 million project would reduce particulate emissions by less than 5 percent in the city, and that it would have negligible effect on the acid rain problem. Environmentalists, including the Polish Ecology Club, argued that U.S. assistance instead be applied to promote fuel-switching to natural gas, elimination of "low-level sources" of particulates, primarily neighborhood-scale boilers, by connecting them to the district heating system, and energy conservation. Approximately $3 million of project funds was allocated to include an energy-efficiency demonstration project. This project demonstrated the energy savings that could be obtained by investing in heat exchangers, heat meters, valves, controls, weatherstripping, radiant reflectors, and insulation. The project has demonstrated a savings potential of 30–40 percent in the buildings.

SOURCE: Tom Secrest, Pacific Northwest National Laboratory, personal communication, 1998.

---

ductions were even smaller than their reduction in GDP, reflecting lack of energy sector reform.

New technology could probably make a difference in the region, but at a considerable cost. The most advanced clean coal system is the Integrated Gasification Combined Cycle system. It couples coal gasification with highly efficient aeroderivative gas turbines (see Figure 5.4 and Chapter 7). This system could emit essentially no sulfur or particulates but would drive the cost of using coal up from $600–700 per kilowatt of installed capacity—including scrubbers—to perhaps twice that much.

Cleaning, or coal washing, can remove pyritic sulfur up to 50 percent in Ukraine, but less elsewhere. Ukrainian coal contains 25–50 percent ash and 2.5 percent sulfur, on average. Coal washing has not been practiced extensively in the East Bloc, but has been recommended as the first and most cost-effective step in reducing emissions. The Czech power com-

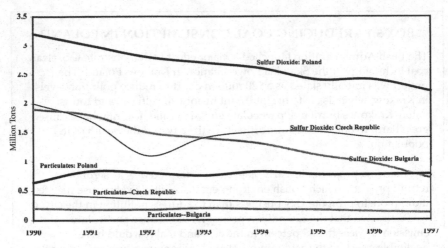

Figure 5.3   Coal-Related Air Pollution in Selected Transition Economies, 1990–1997.
SOURCES: European Environment Agency, "Trends in Acidifying Pollutants,"
http://warehouse.eea.eu.int/cgi-bin/broker.exe; 26 October 1999.

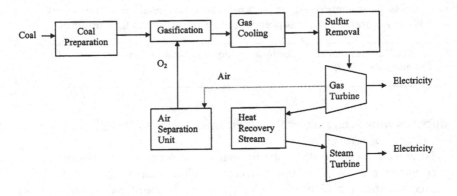

Figure 5.4   Schematic Diagram of an IGCC System

pany CEZ was slated to spend 30 percent of total investment before 2000
on environmental controls for coal plants to meet the requirements of the
Czech Republic's Air Protection Act.

Coal mining in Poland causes serious water pollution. Much of its coal
lies in saltwater deposits. The Ministry of Environment in Poland has

---

BOX 5.2   CZECH REPUBLIC REDUCES
PARTICULATE EMISSIONS BY 70 PERCENT

The Czech Republic managed both to reduce environmental pollution and to avoid deep reductions in GDP. Part of this achievement came as result of conscious policy.

The Czech program for "healing the atmosphere" provided non-business energy users with grants of up to 40 percent of the cost of energy savings measures, and businesses with zero-interest five-year loans for up to 70 percent of the cost of a program. The program was oversubscribed by 50 percent. Forty percent was provided in loans, 60 percent in subsidies. Ninety percent of the program was directed at replacing brown coal use. The savings contributed to a 70 percent reduction in particulate emissions between 1990 and 1997.

SOURCE: SEVEn, The Czech Energy Efficiency Center, Prague, 1998.

---

said that the Vistula does not freeze in winter at Krakow because the salt content is so high, and that in 1995 the Vistula River near Krakow was "saltier than the Baltic Sea."[15] More specifically, the salt concentration of the Vistula 500 kilometers down river at Warsaw exceeds the national standard of 400 mg per liter.[16] In Poland, 65 percent of rivers are unfit even for industrial use, although not just because of brine discharge.

Most of the salt comes from just three large mines owned by the Nadwislanska Coal Company: the Ziemowit, Piast, and Czeczott, which together produce just over 15 million tons of coal each year. The three mines dump 3,500 tons of salt per day in the Wisła River.[17] The economic losses caused by the salt dumping range an estimated $100–250 million per year.[18] At world market prices, the annual value of the three mines that caused the damage by dumping their salt is only $450 million. That is, the environmental externality from the brine dumping of these three mines is equal to at least one-fifth of the revenues of the mines, which almost certainly exceeds their value to the Polish economy.

The salt-dumping problem indicates that neither central planning nor markets alone will solve environmental problems. The history of environmental policy in the United States, for example, is intertwined with efforts to solve what might be described as the failure of tort law. A tort is a damage that one can claim against another person or entity, and it is an idea that evolved from British common law. But if one cannot prove that the molecules of salt that corroded one's pipes or contaminated one's water supply came from a specific mine, plant, or individual, one cannot in the British and U.S. systems collect tort damages. In Poland, as in the United States, any company that acts responsibly and pays extra to remove the poison from a waste stream before dumping it would be disadvantaged by the additional cost in relation to the competition. Unless standards of behavior and pollution are applied uniformly, there is a strong incentive to pollute. Market forces actually discourage the Polish mines from controlling their pollution because doing so would make their coal more costly or even price it out of the market. The mines are already assessed charges for saltwater dumping, which in 1993 amounted to 15 percent of total operating costs. However, it appears that a nonpayments problem has arisen for the mines and their environmental charges. The mines owed over $740 million in fines in 1996.[19]

There are at least four ways to remove salt from the waste steam. These technical fixes include pumping the salty brine back underground; reducing the amount of brine pumped to the surface; and storing the brine for high water level discharge to dilute its effect. Another, possibly more feasible, solution is removal or desalination of the salt from the waste stream. One desalination plant is planned for the Oswiecim area at a cost of just under $400 million. It will reduce the salt dumped into the Vistula by an estimated 40 percent.[20]

Reform

Poland

Poland was the first country to take the painful step of reforming its coal industry. The process involved "corporatization," or the imposition of hard budget constraints. Far from the "shock therapy" of rapid privatization, this moderate step imposed budget discipline without dumping assets onto the market when there were few potential buyers and without allowing the nomenklatura to corrupt the process to make themselves fantastically wealthy, as has been the case in Russia. There were seventy mines in operation, and each of these was required to balance its books. They were combined in 1993 into six joint stock companies and a holding company.[21]

Coal reform in Poland had a severe impact on employment. Polish coal mine employment has gone through a steady, wrenching adjustment. In 1989, some 426,500 workers mined coal in Poland, but their number was reduced to 250,500 by the end of 1997.[22] The Polish government plans elimination of an additional 100,000 coal mining jobs by 2002. Coal output would be reduced from 137 million tons in 1997 to 112 million tons in 2002. A $1 billion World Bank structural adjustment loan was planned to help facilitate the change.[23] Reforming the coal sector is made more difficult than even the numbers indicate by the fact that under the former system, miners ranked among the most highly valued, highly paid workers in the economy. In 1992, miners received 75 percent higher wages than the industry average and were more highly paid than any sector except for nuclear power plant workers.[24] After 1995, miners earned about $75 per month, when they got paid. Hunger strikes, protests, and even blockades have been part of the mining business since the collapse of communism.[25] Overproduction of hard coal in Poland constitutes a severe barrier against alternative fuels use. The present output of coal mines significantly exceeds domestic demand. Export prices are not attractive when compared with the coal extraction costs, meaning the coal will almost certainly be used in Poland.

Ukraine

The World Bank in 1995 concluded that the coal industry in Ukraine could not compete unless it reduced costs by at least 20 percent. The price of domestic coal in the mid-1990s was 20–30 percent above the cost of comparable coal that could be imported. Initially, the Ukrainian government tried to prop up the unprofitable sector by transferring funds from the more-productive to the less-productive mines. These cross-subsidies totaled $400 million in 1995. In addition, the coal industry was exempt from the value-added tax applied to most activities and received grants, credits, and supplements as well. The total subsidy amounted to over 8 percent of the central budget and 4 percent of GDP in 1993. Ukraine recently spent $800 million in subsidies for the coal sector, but despite that fact, coal in Ukraine still costs more than imported coal.[26]

The low productivity of Ukrainian mines translated into low profitability and a pressing need to close mines. Ukraine in 1990 employed 1.2 million workers, over 4 percent of the workforce, some of them—kindergarten teachers for mining communities, for example—not directly involved in production. This number in the first half of the decade fell by about one-seventh, while coal output fell by half. The Ukrainian government in 1995 suggested it would close 39 of 257 mines, far fewer than was needed. The market value of the mines' coal production was about $2 billion, a gross value of only $2,000 per worker. The industry used 6,000 workers to produce 1 million tons of coal, compared to 800–1,500 workers in Western Europe and 100–400 workers in the United States. Even Poland had reduced the number of miners per million tons of production to about 1,800.

Because Ukrainian mines are concentrated in the Donbas region, the social effects will be very severe, as evidenced by a story about a miner immolating himself over $175 in back wages.[27] Restructuring means shutting down twenty mines per year over the next five years, a job that has been assigned to the Ukrainian Coal Company. The cost of providing assistance to workers, including modest retraining, will total $20 million per year, and a World Bank loan of $200–300 million will help facilitate these efforts. Ukraine further suffers from being the manufacturing source of three-fifths of the coal mining equipment for the former Soviet Union, and with a depressed industry region-wide, this sector of the economy also suffers.

The poor condition of the Ukrainian mining sector reflects four problems: poor mine management, redundancy of mine workers, poor geological conditions, and obsolete mining equipment. The first problem theoretically would be addressed by corporatizing and privatizing mines. Eliminating worker redundancies would require closing up to twenty mines per year and laying off perhaps 400,000 additional workers. Nothing can be done about poor geology, of course. Upgrading mining equipment requires capital investment which, like privatization, is hampered by the economic crisis that has driven demand for coal lower than for any other energy source.

Russia

Russian coal mining reform has made some progress in eliminating subsidies and providing hard budget constraints. Budget-based coal subsidies for the industry cost the central government some $2 billion per year in the first half of the 1990s, and although the level of outlay had declined to a billion dollars per year by 1997, a political crisis developed over increasing the subsidy to $1.7 billion. Then-Deputy Prime Minister Chubais objected to the proposal, urging the Ministry of Fuels and Energy instead to increase collections of payments for electric power so that power stations could pay coal mines for the coal they use. A large power plant in Primorskii Krai serving Vladivostok shut down in April 1997 because it ran out of coal. The coal mines could not produce because they had not been paid and could not afford the fuel to transport the coal. Power cuts and blackouts were reportedly lasting twelve hours per day.[28]

Reform in Russia, as in Ukraine, has been made difficult by the large number of obsolete mines and the lack of financing to provide a social safety net for dislocated workers and the social services formerly supplied directly by the coal industry. Russian mines employed over 1 million people. Improvements in Russian mines clearly are overdue. In 1997, some 800,000 miners worked in over 260 mines, half of which were estimated to be unprofitable. Journalist Michael Gordon described a mining operation in the Yuzhnaya mine, nearly 1,000 kilometers (600 miles) south of Moscow:

Wearing padded uniforms and carrying small oxygen tanks that hold an hour and a half's supply of air—a precaution against cave-ins—miners crowd into a shaky elevator for the long descent. . . . After a jarring train ride through miles of tunnels, another steep descent in a second train and a march through pools of dank water and clouds of dust, the miners begin six-hour shifts, scraping coal out of deep, but narrow, crawl spaces. . . . They do it for the same reasons that their fathers did it. It is practically the only job in town.[29]

The World Bank in June 1996 approved a Coal Sector Adjustment Loan (Coal SECAL). The funds went to pay for housing and utilities in coal mining communities in each of the three major coal basins, Kuzbas, Rostov, and Tula. Russia in return pledged to transfer management of the states' shares in the coal industry to a private trust that would provide professional management for two-fifths of the industry. The loan totaled $500 million and was provided in two tranches in 1996. The second tranche was delivered in December when a World Bank mission determined that the industry had insufficient funds to provide heating and housing for the winter.[30]

The World Bank in 1997 proposed a second coal sector structural adjustment loan, citing considerable progress. Bank staff noted that between 1993 and 1997 subsidies for coal had fallen from 1 percent to 0.2 percent of GDP. Fifty-eight mines were closed, and the sector divested its social assets such as housing and kindergartens.[31]

By 1998, 200 mines in Russia remained in operation, with 86 scheduled for immediate closure. Some 50,000 miners were to be laid off, and half the coal sector was to be privatized in 1998, and this policy continues to be implemented. The government budget allocated about $1 billion to the coal industry in 1998, half the amount said needed by the Independent Union of Coal Industry Workers. Two-thirds of the budgeted amount was to support the closure of unprofitable mines.

Conclusions

The coal sector in Eastern Europe and the former Soviet Union faces the grimmest reform task. Because coal is difficult to mine, transport, and burn, demand for the fuel in the transition economies has fallen more than for any other energy source. Because workers in the sector face terrible conditions, obsolete equipment, and weak management, their numbers have been and must continue to be cut drastically.

The west, especially the development banks, can help mitigate this pain. Key nations in transition have already taken the painful step of eliminating almost 1 million jobs in coal mines. They have pushed coal prices throughout the region to world market levels, slashed subsidies by half and more, fired workers, and divested the social assets and spun off social services from the coal mining sector.

Serious structural challenges remain. Coal mining technology seriously lags behind the western world, causing severe productivity problems and catastrophic loss of life. Pollution control equipment, even coal washing, is often lacking in coal-burning systems, creating ongoing health and environmental problems. Increasing unemployment in the coal mining sector has already reached levels that create social tension, and large, additional displacement of workers is inevitable.

This type of reform was bound to be the most difficult, and promises of western assistance have never been commensurate with either the need or the verbal commitments made at the time of the collapse of the Soviet Union. An important exception to this general failure of western policy is the structural adjustment lending provided by the World Bank to aid in closing down unnecessary parts of the coal industry. This assistance has been criticized for not having reached the miners, criticism that is not altogether fair because much of the funding went to governments to help finance the transition. The west will benefit environmentally from the reduction of carbon dioxide emissions and—at least in parts of Europe—of sulfur dioxide emissions. However, the coal sector will need large-scale assistance for the next five to ten years, if the region is to end the ongoing tragedy of coal mining accidents, coal-related pollution, and severe unemployment.

Notes

1. Pablo Neruda, "The Central Hand," in *Elemental Odes,* translated by Carlos Lozano (New York: G. Massa, 1961).

2. A methane blast in a coal mine in Donetsk killed 39, raising the death total in Ukrainian coal mines during the first six months of 1999 to 160. "Methane Blast in Donetsk Kills 39 Miners," *RFE/RL Newsline* (25 May 1999); http://www.friends-partners.org/friends/news/omri/1999/05/990525I.html.

3. Conference on Environmental Health in Central and Eastern Europe, John Lampe, chair, Woodrow Wilson Center, Smithsonian Institution (Washington, D.C., 1989).

4. *Russian Energy Picture,* CENEf (Moscow, January-March 1998), 44.

5. Robert E. Ebel, *Energy Choices in the Near Abroad: The Haves and Have-nots Face the Future* (Washington, D.C.: Center for Strategic and International Studies, 1997), 81.

6. Office of Technology Assessment, *Fueling Reform: Energy Technologies for the Former East Bloc* (Washington, D.C.: Government Printing Office, 1994), 107.

7. William U. Chandler, *The Changing Role of the Market in the World's Economies* (Washington, D.C.: Worldwatch Institute, 1986).

8. World Bank, *Ukraine Coal Industry Restructuring Sector Report*, World Bank Staff Report (Washington, D.C., March 1996), iv.

9. International Energy Agency, *Energy Policies of the Russian Federation* (Paris: Organisation for Economic Cooperation and Development, 1995); BPAmoco, "Statistical Review of World Energy," http://www.bpamoco.com/worldenergy/; British Petroleum, *Annual Statistical Review, 1998*, www.bp.com.

10. Stephanie Baker-Said, "Paid Off Then Laid Off," *Moscow Times*, 21 April 1998, p. 3.

11. International Energy Agency, *Energy Policies of Poland* (Paris: Organisation for Economic Cooperation and Development, 1994), 183.

12. World Bank, *Environmental Action Programme for Central and Eastern Europe: Setting Priorities* (abridged version of the document endorsed by the Ministerial Conference, Lucerne, Switzerland, 28–30 April 1993). See also Clyde Hertzman, *Environment and Health in Central and Eastern Europe*, report for the Environmental Action Programme for Central and Eastern Europe, World Bank (Washington, D.C., 1995).

13. Aleksandr Avdiushin et al., *No-Regrets Options in Climate Change Mitigation Policy: Lessons from Transition Economies*, Battelle, Pacific Northwest National Laboratories (Richland, Wash., May 1997); U.S. Energy Information Administration, "Poland: Environmental Issues," http://www.eia.doe.gov/cabs/polenv.html#AIR-POLLUTION, March 2000.

14. IEA, *Energy Policies of Poland*, 183.

15. Paul Wrabec, "Kopalne przed trybunalen . . .," *Gazeta Wyborcza* no. 198 (25 August 1993): 15. See also Maciej Nowicki, *Environment in Poland* (Warsaw: Ministry of Environmental Protection, Natural Resources, and Forestry, 1992), 13; and "Environment Ministry Reports on Water Pollution," *Rzeczpospolita* no. 127 (2 June 1995): 2.

16. Jerzy Kurbiel, Wojciech Balcerzak, Stanislaw M. Rybicki, and Kazimierz Swist, "Selection of the Best Desalination Technology for Highly Saline Drainage Water from Coal Mines in Southern Poland," *Desalination* 106, nos. 1–3 (August 1996): 415–418.

17. Thanks to Patricia Mitchell for her analysis of this issue and for calling sources to my attention in "Possibilities for Wind Energy in Ukraine" (in partial fulfillment of master's degree requirements, Johns Hopkins University School of Advanced International Study, Washington, D.C., May 1997). See also John Chadwick, "Nadwislanska and Rybnicka," *Mining Magazine* (March 1996).

18. Marian Turek, Julita Mrowiec-Bialon, and Witold Gnot, "Utilization of Coal Mine Brines in the Chlorine Production Process," *Desalination* 101, no. 1 (March 1995).

19. "Environmental Problems Facing Area of Heavy Industry in Poland," *Rzeczpospolita* (26 July 1996): 19. The sum of 2 billion 1996 new Polish zloty was converted at a rate of 2.6954 per U.S. dollar based on data provided by Sławomir Pasierb, Polish Foundation for Energy Efficiency, personal communication, Katowice, Poland, October 1998.

20. "Mine Water Desalination Plant to Be Built in Oswiecim," *Rzeczpospolita* (3 April 1996): 2.

21. International Energy Agency, *Energy Policies of Poland: 1994 Survey* (Paris: Organisation for Economic Cooperation and Development, 1995), 86.

22. For 1989 data, see *Charakterystyki Energo-Ekonomiczne Galezi i Branz Przemyslu*, Centrum Informatyki Energetyk (Warsaw, Poland, 1992). For 1997 data, see "Projekt Reformy Gornictwa" (special section), *Gazeta Wyborcza*, 12 April 1998; Darius Malinowsky, "Program restruktury zacji gornictwa wtorek gorikow," 11–13 April 1998, p. 24.

23. Christopher Bobinski, "World Bank to Back Polish Coal Plans," *Financial Times*, 12 June 1998, p. 3. See also "World Bank Resumes Lending for Coal Sector and Social Protection in Russia," World Bank press release, Washington, D.C., 23 July 1999.

24. See generally Energy Sector Management Programme, *Poland—Energy Sector Restructuring Program*, Volume 2, *The Hard Coal Sector*, Report No. 153/93 World Bank (Washington, D.C.).

25. Peggy Morvant, "Tradegy in Kuzbass," *OMRI Daily Digest* II, no. 172 (5 September 1995); http://www.friends-partners.org/friends/news/omri/1995/950905I.html.

26. Meredydd Evans, Battelle, Pacific Northwest National Laboratory, personal communication, Washington, D.C., 24 January 1997.

27. See, for example, "Ukrainian Miner Immolates Himself Over Unpaid Wages," *RFE/RL Newsline* 3, no. 16, pt. II (25 January 1999); http://www.friends-partners.org/friends/news/omri/1999/01/990125II.html.

28. "Non-Payments Lead to Massive Power Cuts in Primore," *RFE/RL Newsline*, (4 April 1997); http://www.rferl.org/newsline/1997/04/040497.html.

29. Michael R. Gordon, "Russia to Take Scalpel to Coal Industry," *International Herald Tribune*, 1 March 1996, p. 13.

30. "World Bank Helps Reform Russia's Coal Industry," World Bank press release, Washington, D.C., 23 December 1996.

31. Janet Koch, Europe and Central Asia Department, *Russia-Coal Sector Adjustment Loan II*, Report No. PID5822, World Bank (Washington, D.C., 29 October 1997).

6

Utility Monopoly and Competition

When the transition economies and their western consultants began to privatize and restructure their national utility monopolies in the 1990s, they joined a growing worldwide trend. Going beyond mere privatization, they attempted to create competitive power markets in a sector that from its creation had everywhere been a regulated monopoly. The model they copied was Great Britain's, with limited examples from certain U.S. states.

Creating competition in the transition economies faced significant problems. First was a general lack of experience with privatization in the region and with competition elsewhere. European and U.S. competition policy has evolved very slowly. Indeed, a large share of the U.S. power supply sector remains government-owned.[1] Even British power generation comprises just three companies, one of which is subsidized by the others' power sales.

Second, retooling the region's power systems for efficiency and environmental controls and replacing worn out capital could cost more than $100 billion. Large amounts of capital would need to come from outside the region, raising local concerns about ceding control of a strategic sector to foreigners. Dumping so much capital onto the market could also mean giving it away at fire-sale prices. Competition would also require expensive computer and control equipment, a requirement with which even Great Britain has been struggling.

Third, Central European nations face the additional complication of convergence with the European Union, where reform is just getting underway. In 1996, European Union energy ministers, meeting in Luxembourg, agreed to dismantle vertically integrated utilities in member countries and permit location of suppliers of power for any country in any country. However, the initial share of each national market that would be required to be open to international competition would total only about one-fifth of the market, rising only to one-third by 2006.[2] Integration is made difficult by the need to standardize systems; consensus on even seemingly small matters such as a standard size for electrical outlets has not been reached.[3] Finally, 47 gigawatts of nuclear power spread across nine countries creates public versus private conflicts regarding liability for reactor accidents, nuclear waste, and power plant decommissioning, issues that elsewhere have taken decades to resolve (see Table 6.1).

It is striking that despite these challenges, Hungary has privatized its electric utility sector, Ukraine has attempted to create a competitive market for power at the wholesale level,[4] and Poland has partially restructured and commercialized its power sector. Nevertheless, utility reform lags behind that of almost every other sector of the transition economies.

Highly Charged Reform

Central European utility reform has been heavily influenced by Great Britain, which has created limited competition in generation and full competition at the retail level. Three British companies generate power, including a subsidized, heavily regulated firm operating Britain's nuclear plants. At the retail level, fourteen retailers, or distribution companies, compete for each others' local customers. The companies compete to acquire power from the generators, and may sell this power to power users over each others' lines. That is, the distribution system serves as a common carrier. This competition puts price pressure on service providers and some pressure on the generators. Critics have argued that there is a role for public regulation of both prices and profits for utilities during at least the transition to a more competitive market. They have cited, for example, delays in the installation of computer and control equipment necessary for registering transactions.[5] In late 1997, British Energy Minister John Battle, saying that he had become concerned about the security of power supplies when he read in a newspaper that coal production was threatened by natural gas-fired power generation, issued a moratorium on permits for new gas-fired capacity.[6] The government later planned to extend this moratorium for five years.[7] These technical and regulatory issues will affect the transition economies, particularly the tradeoffs with unemployment in the coal sector.[8]

The transition economies have also been influenced by U.S. policy changes affecting competition. The U.S. federal government endorsed electricity competition in a plan announced in 1998 that would fully take effect by 2003.[9] Competing U.S. interests, however, have slowed reform. The U.S. interstate wholesale power market could eventually be worth

TABLE 6.1   Installed Power Capacity by Source, Selected Transition Economies, 1998 (Gigawatts)

Country	Fossil	Hydro	Nuclear	Total
Bulgaria	7	1	4	12
Czech Republic[a]	11	1	2	14
Hungary	6	0	2	8
Poland	27	2	0	29
Romania	16	6	1	23
Slovakia[a]	3	2	2	7
Kazakhstan	17	2	0	19
Russia	146	44	21	211
Ukraine	37	5	13	54
Total Region[b]	308	81	47	435

NOTES: a) Totals have been rounded. b) Nineteen countries. Note that values may differ from official national statistics.

SOURCE: Energy Information Administration, U.S. Department of Energy, "International Energy Information," http://www.eia.doe.gov/emeu/international/contents.html, May 2000.

over $45 billion, providing impetus to competition. However, existing utility monopolies have protested that they are owed compensation for "stranded investment." This claim stems from the assumed social contract between private investors who guaranteed that customers would always have power available in return for monopoly control and a guaranteed return on capital investment.[10] In addition, utility regulation had to some degree made provision for environmental and social externalities by providing preferences for "clean energy" and for "life-line" power rates to protect poor consumers. Some critics of deregulation argue that these values will not be provided for in a purely competitive market.

Privatization of utilities—natural monopolies—requires special steps. According to the World Bank, these include:[11]

1. Determining whether parts of the system can be competitive, and separating those and introducing competition.
2. Creating the legal and institutional framework to regulate the natural monopoly, including setting prices and rates of return on investment.

3. Commercializing—introducing hard budget constraints for—all power sector enterprises.
4. Attracting private investment through concessions or privatization.

Early efforts in utility reform in Poland were initiated by nongovernmental organizations, including the American Council for an Energy-Efficient Economy, and bilateral agencies, including the British Know-How Fund with cooperation of the U.S. Agency for International Development.[12] The Polish approach assumed that electric power could be provided most efficiently, in an economic sense, if competition were created to generate the power. The idea was to break the vertically integrated, state-owned electric utility into three parts. Power generating stations forming the first part were commercialized as more than thirty private firms competing to sell wholesale electric power. Commercialization of the electric power sector was to be handled initially by the Ministry of Ownership Transformations and later by the Ministry of Treasury. A few power plants were privatized. The second part of the system, the national power grid consisting of high-voltage transmission lines and the power dispatch operations, was corporatized as the Polish Power Grid Company, but not privatized. The Polish government retained ownership of the Grid Company, and regulates its basic operation. Power generators partially compete to sell power to the Grid Company, which purchases power for the best price. A significant part of power market competition was to long- and medium-term contracts to encourage modernization of power plants, including installation of environmental controls.

The local power distribution companies form the third part of the restructured system. Some thirty-three firms have been separated from the national monopoly as state-owned joint stock companies. They were given exclusive power distribution rights in their territories, at least in the near term, for power supply up to 110 kilovolt networks serving residential, commercial, and industrial customers.

The plan is being implemented in three phases over the fifteen-year period between 1993 and 2006. The first phase, completed in 1997, provided for long-term contracts-four years in length-between the Polish Power Grid Company and generators. Distribution companies simply bought power at cost and maintained their monopoly of local service. Combined heating and power plants were allowed to sell to distributors at the same price offered by the Grid Company to the distributors. Any savings were allocated to district heating customers. In the second phase, starting in 2000, competition for the pool market will be created among generators bidding to sell power to the Grid Company. The philosophy of the new

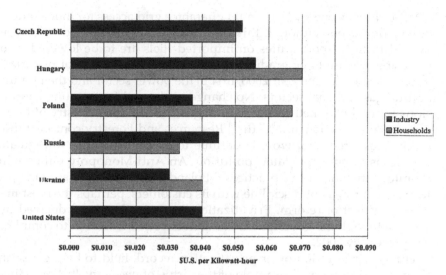

Figure 6.1 Electricity Prices in Selected Countries, 1998.

SOURCE: *Energy Prices and Taxes—Quarterly Statistics (Fourth Quarter 1998)*, Part III, Section B, Table 14 (Non-OECD Countries) (Paris: International Energy Agency, 1999).

energy law, enacted in 1997, is to create a fully competitive energy market. Third-party access and transmission service was opened in 1999 for large buyers and sellers of electricity and district heat. Beyond 2005, full competition, including third-party access to the grid for independent power producers, will be allowed, and contracts will be made directly between distributors and generators and smaller customers and distributors. In technical terms, retail wheeling will be permitted after 2005.[13]

Pricing is a critical index of utility reform. Polish industrial and residential electricity prices by early 1997 were comparable with those of the United States (see Figure 6.1). Significantly, residential electricity was priced higher than industrial electricity. Since 1999, electricity prices have fully reflected the short-term cost of generation and distribution for all customers. The Energy Regulatory Authority, created by the energy law, regulates electricity prices transparently on the basis of published rules. Despite that fact, some transition economies still have "inverted rates," meaning that industrial power is priced higher than residential electricity, which costs much less to deliver. This inversion typically indicates a cross-subsidy, from industry to households. The Polish rate structure, however, encourages efficiency by placing the burden of cost where it is incurred, and also has positive economic and employment benefits.[14]

Poland's new energy law would guarantee grid access for independent power producers, although this right for gas will be limited to Polish-owned firms. Import duties on imported fuels are to be lowered, and price controls are to be gradually phased-out. Acceleration of privatization is planned for the year 2000, first in the power sector and then for the natural gas and coal sectors. No change is planned for granting concessions for coal bed methane or natural gas and oil. The Ministry of Environmental Protection and Natural Resources and Forestry controls those natural resources and would continue to do so, as well as to regulate emissions of energy-related pollution. An Anti-Monopoly Office will monitor anti-competitive practices.[15] Poland's power sector has managed to maintain payment discipline with its customers, perhaps the most important aspect of reform. Privatization, however, has been delayed by years, although the government has announced its intention to complete the process by 2002.

Hungary's electric power privatization record, held to be the best in the region, is a premier example of the virtue of necessity. In 1997, Hungary became the first transition economy to privatize power generation by selling off all its power generating capacity, except its nuclear plants. A driving force behind the privatization of the Hungarian energy system was a budget deficit, which had exceeded 8 percent in 1994. The nation had a much more severe debt burden at that time than the rest of Central Europe and needed to generate revenues.[16] Hungary's Electricity Law of 1994 broke the vertically integrated monopoly into separate generating companies, a national grid company, and local distribution companies. The grid and distribution companies retained monopoly control over power transmission and distribution. The power generation business was opened to all suppliers, while the grid company serves as the buyer for power from all plants larger than 20 megawatts. The Privatization Law of 1995 permitted the sale of generation and distribution companies as well as the grid. Open, competitive bids returned over $1.25 billion to the Hungarian treasury, plus commitments for investments by buyers in the power plants themselves. Sales to bidders were limited to 50 percent of shares, and they were all purchased by foreign companies, mainly from Germany, France, and Britain. A portion of the remaining shares was distributed to municipalities, employees, and holders of Hungarian privatization vouchers. The grid company, however, could not be sold for lack of a bidder. Prices had by 1997 reached levels high enough to ensure cost-recovery.[17] Moreover, the cross-subsidy from industries to residences had been eliminated.

Hungary's privatization was delayed by fights over jobs, pricing, and nuclear energy.[18] The electrical workers' trade union fought the move to protect their jobs, and aligning nuclear power with markets was also a

TABLE 6.2   Power Plant Utilization in Selected Transition Economies, 1997 (Capacity Factor)

Country	Percent
Bulgaria	37
Czech Republic	46
Hungary	51
Kazakhstan	32
Poland	48
Romania	28
Russia	39
Ukraine	35

NOTE: Capacity factor is actual power generation divided by maximum possible (installed capacity times 8,760 hours per year). The U.S. value in 1998 was 49 percent.

SOURCE: Author's estimates; Energy Information Administration, U.S. Department of Energy, "International Energy Information," http://www.eia.doe.gov/emeu/international/contents.html, May 2000.

challenge for Britain, where a cross-subsidy on non-nuclear power consumers was imposed. Hungary, however, enjoyed two key advantages over other transition nations in its power sector that made it more likely to succeed in early privatization of power generation. First, power supply was more closely matched with demand, meaning that the market for power sales was stronger. The capacity factor—the percent of time power plants actually generated power—was at the time that Hungary's system was sold the highest in the region. Hungarian power plants operated an average of 51 percent of the time, a value comparable to the U.S., German, and French experience. A power company owning a newly privatized plant would have a much higher probability of being able to sell that power in Hungary than in, say, Romania, where the ratio was only 30 percent, meaning that power plants would on average be idle more than 70 percent of the time[19] (see Table 6.2).

Second, Hungary's power privatization also had the advantage of a relatively strong national economy. Hungary's GDP grew slowly during the period—1.5 percent in 1995, 1 percent in 1996, and 2.5 percent in 1997—but positive figures stood in marked contrast to experience in the larger power markets of Russia and Ukraine.[20] Hungary also benefited from a 1997 World Bank loan of $60 million to improve grid reliability.

The loan enabled planners to add a gas-fired power plant, which could be switched on quickly to meet peak demand and to maintain power quality.

Czech energy analyst Jirka Zeman describes the Czech utility sector at the end of the 1990s as still based on the state-owned enterprise model. In principle, competition from independent power producers was permitted in the Czech Republic. In reality, CEZ, the national power generating company—two-thirds state-owned—dominated the sector, controlling 80 percent of generation.[21] Electric power prices reflected the extent of government intervention. Prices were inverted, with industry paying almost 60 percent more per kilowatt-hour than households. Prices have been regulated by the Ministry of Finance since 1996.[22] However, foreign interests have bought heavily into gas and electric distribution utilities, and the Czech government in 1999 pledged to accelerate privatization.[23]

Czech policy is expected to continue to support clean air, natural gas, and a heavy reliance on nuclear. The government has articulated key utility-related policies:

- Reducing dependence on coal
- Completing the Temelin nuclear power plant
- Keeping the Dukovany nuclear power plant in operation
- Closing one-quarter of coal-fired power capacity
- Installing flue gas desulfurization scrubbers on 6,000 megawatts of power capacity
- Meeting European power quality requirements to connect to the European grid

Since 1994, the Czech utility has invested several hundred million dollars of additional funds to upgrade the Temelin nuclear plant to western standards. Although the plant was expected to be completed in 1998, it is now not expected to be available until 2005.

The decision by CEZ to complete the Temelin nuclear power station has added a wrinkle to Czech power reform. The plant, four-fifths complete at the time of the Velvet Revolution, would provide 1.9 gigawatts of capacity from two new VVER-1000 reactors retrofitted with Westinghouse control systems. Completion was made possible by U.S. Ex-Im Bank financing. Temelin's pending completion, however, has created the expectation of an over-supply of power, thus delaying alternative investment and even prompting CEZ to promote electric resistance home heating. Meanwhile, the plant's completion has been delayed several times and may not be realized before 2005. If completed, nuclear power would provide one-third of Czech electricity, and coal would fall to one-half.

Ukraine, by all rights, should be top ranked in utility reform. Creating the first wholesale market for power in the region in March 1997, the nation stood at the global cutting edge of utility reform. However, wholesale competition in the Ukrainian utility sector, according to World Bank economist Laszlo Lovei, has been stymied by state ownership. Lovei noted that "If you all have the same owner you cannot be truly competitive."[24]

The Ukrainian utility industry was broken up in 1996, with the vertically integrated power generation, transmission, and distribution system transformed into parts with separate joint stock companies at the generation level, very much as in Poland. The generating companies, for example Donbasenergo, offered hourly bids to a central agent at the wholesale level. The generating companies were to be offered to private investors in September 1997.[25] Competition was introduced at the same time, along with trading of wholesale power in April 1996. A state-owned enterprise, EnergoMarket, made the market and managed the bidding, selection of prices, and payments to generators. A National Electricity Regulatory Commission was responsible for licensing power suppliers and protecting consumers. A World Bank Power Sector Development Loan was instrumental in establishing this market.[26] This system collapsed under the weight of delinquent bill payments.

One major problem was lack of time-of-day metering for charging for variable electricity rates. Another was that nuclear power did not compete, but received a fixed price.[27] Ukrainian power sector reform was marked by crisis in 1997 when the World Bank suspended its $317 million power sector structural adjustment loan to the country. The financing, like all structural adjustment loans, provided the national government with resources necessary to adapt the power sector to market conditions, not to buy capital equipment. This particular loan was meant to help pay for fuel, spare parts, and the installation of electric meters. The World Bank halted disbursements of the loan when Ukraine failed to raise power prices on the schedule to which it had agreed. Delay and controversy in power sector reform continued in 2000.[28]

The Ukrainian government attempted to bring discipline to the power sector by threatening to cut off power to nonpaying customers. A Ukrainian Financial Recovery Plan agreed to mandate disconnection of service for nonpayment, to collect debts, to promote privatization, and to meet a timetable for these and other reforms. The government also insisted—per Cabinet of Ministers decision—that barter payments had to be made in accordance with transparent banking practices defined by the Bank of Ukraine.[29] But this critical reform was not enforced with sufficient consistency and rigor to overcome the nonpayments problem. Pricing policy has also been uneven. Prices were increased to replacement levels, but with reductions and exceptions later allowed, especially for cogenerated power.[30]

Russia's problems with utility reform are much larger, literally, because its power system stretches halfway around the world. At stake in Russia are two key blocks of power. The first is the share of thermal, or fossil, fired generating capacity of about 80 gigawatts that might be retired by the year 2010. Half of that capacity is provided by combined heat and power stations, also called cogeneration plants or "TETS." The other block at stake is 20 gigawatts of nuclear capacity generally considered to be unsafe. A key energy-environmental question for the next quarter century is whether that capacity can be replaced by low-emissions, high-efficiency combined cycle gas-fired power plants.[31]

Russia's Unified Electrical Power System (RAO EES Rossiya), a holding company, controls 70 percent of Russia's electric power generation system.[32] The company in 1995 controlled 212 gigawatts of capacity at over 700 generating units, and another 43 gigawatts owned by the Central Dispatch Office. EES Rossiya also controls the power grid distribution system. Moreover, it effectively controls the regional "energos," or power companies, which nominally own 135 gigawatts of power generating capacity.[33] Although RAO EES Rossiya is owned by stock holders, the Russian government holds and intends to maintain ownership of 49 percent. That, plus the fact that the company's president is appointed by the government, leaves effective control of the company in the hands of the central government.

Russia's transmission system is the largest in the world, but inter-ties between some of the internal grids are inadequate. This inadequacy creates two complications for closing unsafe nuclear reactors. First, the need to shut down the reactors is made greater because lack of reliable connections means that backup power for emergency cooling is always in serious jeopardy. In this regard, Russia does not even meet the most basic world standard of being able to lose generating units without having to resort to blackouts or brownouts. Second, power from the Kursk and the Leningrad stations is connected in regions with particularly weak inter-ties. This weakness makes closing those particular plants more difficult. Building the necessary inter-ties would cost less than $800 million. The cost of stabilizing the entire transmission system, however, is probably on the order of $9 billion. Eighty percent of this cost would be incurred for better integrating Siberian plants and far eastern hydroelectric capacity.[34]

Russian industrial prices are quite high by regional standards, a difference caused in part by a cross-subsidy to households. This practice was scheduled for elimination by 2003, according to a government decree. In the aftermath of the financial collapse and the subsequent political retrenchment, this reform has been put on hold.[35] Inefficiency is great as well, with older, gas-fired plants averaging 30 percent or even less. New thermal stations convert only 35 percent of fuel to electricity, whereas advanced combined-cycle gas turbines, which have provided most new electric power capacity in the United States for the last decade, can cost-effectively exceed 45 percent. Russia, with its vast gas resources, would seem perfectly suited for a competitive electric power generation system.

A serious problem is the division of responsibility for the sale of power from nuclear power plants and the responsibility for maintaining and operating those plants. Currently, EES Rossiya simply receives the power from the joint stock company Rosenergoatom, which is wholly owned by the Ministry of Atomic Energy (Minatom), and transmits it to customers. But because EES Rossiya suffers a massive nonpayment problem from its customers, Rosenergoatom in turn does not get paid. This situation led to dangerous confrontations at nuclear facilities when workers threatened strikes and shutting down reactors, seizing one power station control room.[36] As a result, the nuclear generator cannot ensure that it gets paid enough to maintain the safety of its plants, and EES Rossiya lacks authority to determine its most cost-effective sources for power.[37]

Russian utility reform has been bogged down at a much more fundamental level. An illustrative battle pitted reformers, including former Deputy Prime Minister Boris Nemtsov, against local power company bosses in St. Petersburg and Krasnoyarsk. This struggle followed the inauguration of a wholesale power market the previous month. Local utilities had opposed the market because they profited from their monopoly position and because local governments wanted to maintain residential power subsidies made possible by cross-subsidies charged to large, local enterprises that had no choice under monopoly conditions.[38] Even earlier, United Energy Systems and the reform government had forced Mosenergo, which provides power and heat to Moscow, to abandon plans to limit shareholders' rights. The move had been particularly targeted at foreign investors. The EBRD apparently helped pressure Mosenergo to give in.[39] When Nemtsov managed to oust bosses from those regions and to remove some two-thirds of the old board of directors of the national utility monopoly, United Energy Systems, the company's stock increased in value almost by half.[40] Again, in 1999 when Anatoly Chubais was installed as head of RAO EES without the consent of shareholders, observers charged that the Kremlin was attempting to control the system to influence pending elections.[41]

Nations such as Poland and the Czech Republic have reached the stage where they may solve chronic management problems in the utility sector. Ukraine and Russia have not (see Table 6.3). Concerns about the legal

TABLE 6.3 Status of Electric Utility Reform in Selected Countries, 1999

Country	Structure[a]	Ownership[b]	Regulatory Authority
Bulgaria	Vertical monopoly	State enterprises	Ministry of Energy & Energy Resources
Czech Republic	Flat monopolies	State joint stock	Ministry of Industry and Trade
Kazakhstan	Flat	Mixed joint stock	Kazenergo
Hungary	Flat, competitive[c]	Private	Energy Office/Ministry of Industry
Poland	Flat, competitive[c]	State joint stock	Energy Regulatory Authority
Romania	Vertical monopoly	State	Ministry of Industry
Russia	Vertical monopolies	State joint stock	Federal Regulatory Commission
Slovakia	Vertical monopoly	State	Ministry of Industry
Ukraine	Flat	State	National Regulatory Commission

NOTES: a) "Flat" implies separation of generation, transmission, and distribution systems. b) Joint stock is shorthand for joint stock companies. c) Wholesale competition only.

SOURCE: John A. Riggs, Rapporteur, "Utility Restructuring in Central and Eastern Europe," (report of the 2nd Central and Eastern European Energy Workshop, 7–10 May 1997, Krakow, Poland, Aspen Institute, Washington, 1997); updated by the author.

foundations of utility reform have long been the top priority, to stop the massive loss of revenues by enforcing payments for utility services. Some experts argue that private companies can best do that because often it is a problem simply of sending out bills and following up on payments. They note that as long as utilities have ties to governments, they will be forced to carry out social or political objectives such as making sure that citizens do not freeze in the dark or that factories have sufficient power to run to provide jobs in towns—and in Russia, often whole cities—that have no alternatives whatsoever.[42] Some of those objectives are worthy and will not go away with privatization. The former Soviet region would seem to enjoy an advantage over western countries, with a milder "stranded costs" problem, partly because so many plants are worn out (see Table 6.4).[43] The obsolescence of several tens of gigawatts of generating capacity, which usually lacks environmental controls, has prompted some pollution-plagued Central European countries to consider utility taxes or pollution fees to fund energy efficiency programs and low-income assistance. Such provisions have generally been weak or nonexistent, however.

Transferring ownership of a monopoly from the state to private capitalists just substitutes one set of problems for another, especially if no regulatory mechanism is in place. However, with that framework for regulating the functioning natural monopoly and with the natural monopoly fully commercialized, competition in other parts of the system and private investment in them makes sense. With the exception of Hungary, the region has not accomplished these first priorities.

District Heating

Direct and indirect energy costs consume 25–35 percent of municipal budgets in many Russian cities. Heat subsidies alone make up 15–25 percent of municipal expenditures. Some 1,500 households were left without heat in Murmansk in subfreezing weather in December 1996 when a gas utility refused to provide gas to a city utility, which was in arrears on gas payments.[44]

Improving energy efficiency, especially in district heating systems, is therefore a top priority for local and regional governments (see Box 6.1). Housing and public utility services reform has finally become a focus of social and economic policies. In his 1997 annual message to the Federal Council, Russian President Boris Yeltsin highlighted this reform as a major economic problem. Public funding for this sector amounts to 100 trillion rubles, or $ 17.5 billion, annually. Shifting this burden to residents would probably be technically impractical and politically impossible, given current Russian finances. There are other possible options, how-

TABLE 6.4   Electric Power Capacity Obsolescence, Selected Countries

Country	Obsolete by 2005 (megawatts)
Czech Republic	6,000
Hungary	3,000
Poland	2,800
Russia	50,000
Ukraine	1,400

SOURCE: International Energy Agency, *Energy Policies of the Russian Federation* (Paris: Organisation for Economic Cooperation and Development, 1995); International Energy Agency, *Energy Policies of Ukraine: 1996 Survey* (Paris: Organisation for Economic Cooperation and Development, 1996); International Energy Agency, *Energy Policies of Poland* (Paris: Organisation for Economic Cooperation and Development, 1995); International Energy Agency, *Energy Policies of the Czech Republic: 1994 Survey* (Paris: OECD/IEA, 1994); International Energy Agency, *Energy Policies of Hungary: 1995 Survey* (Paris: Organisation for Economic Cooperation and Development, 1995).

ever, such as eliminating public heating monopolies and shifting subsidies to finance efficiency improvements so that the fuel subsidies can be reduced or eliminated.

Russian and Ukrainian law impedes the installation of distributed boilers, those boilers for which residents would pay directly through a cooperative or building owner. Distributed boilers would ensure more reliable heating, preventing the frequent, sometimes extended periods of heat and hot water loss. Repealing such codes and ordinances would be essential for switching consumers from subsidized central heating to pay-as-you-go distributed heating. Alternatively, corporatization and privatization may provide the incentive for utilities to collect heating payments. Utilities would have to be given the authority to refuse provision of heat for nonpayment; however, this is a politically explosive option.

The power of local utilities to overcharge customers—even when they were their municipal owners—has been demonstrated in the district heating sector. According to Russian economist Igor Bashmakov, contracted amounts of energy have often been overstated in comparison with the amount of heat delivered. In a similar case, in Kostroma, the city was credited for residential customers' payments for only one-twelfth the amount actually paid. The utility had taken a single monthly payment—payments are in equal amounts—and failed to multiply by the number of months in a year. The city had no idea it was being over-

---

BOX 6.1   HEAT OF RUSSIA PROGRAM

The District Heating Initiative or Heat of Russia Program was introduced to increase the efficiency and reliability of Russian district heating systems. The three main objectives of the program are to develop a viable customer base for efficiency upgrades, create a financing stream from the diversion of heating subsidies to efficiency investments, and ensure a market share for private companies. To implement the program, leaders focused on the progressive cities of Chelyabinsk, Lytkarino (outside of Moscow), Dzerzhinsk (a district of Nizhny Novgorod), Krasnoyarsk, Kostroma, and most recently Murmansk.

Each city has committed to shift subsidies from supply to finance energy efficiency upgrades in their district heating systems. Business plans have been completed for Chelyabinsk, Lytkarino and Dzherzhinsk, Kostroma, and other cities.

SOURCE: Igor Bashmakov, Center for Energy Efficiency, personal communication, Moscow, June 1999.

---

charged until the books were audited by CENEf, an independent organization.[45]

Non-payments for heat to Russian regional governments at the end of 1997 ranged from about 50 percent in Dagestan and Moscow to more than 85 percent in Komi-Permyak and the Jewish Autonomous Oblasts.[46]

Czech city-owned distribution companies similarly incur significant costs—almost $9 per gigajoule—for providing district heat. But as a matter of federal regulation, not unrelated to various elections, the utilities charge their customers $2.50 less than costs, which amounts to a quarter billion dollars per year in Czech government subsidies.

Generally, Czech energy policy is oriented to achieving the goal of joining the European Union.[47] Coal will be phased out for district and local heating. An example of how environmental penalties and market economics work together is in the Northern Bohemian city of Most. The Ústí nad Labem city formed the Ústí Cogeneration Plant company from a state-owned enterprise to provide heat and power to the city. The newly privatized firm, facing huge fines from air pollution emissions, invested in desulfurization, turbine upgrades, and electronic control technology. Installation of the pollution control equipment began in 1994. It will reduce local NO_x concentration by 55 percent. Total cost of the project will be $69 million. Twenty percent of the cost of the project will be financed with a long-term loan from the Czech Savings Bank, and the rest with sales of bonds. The turbines will be leased. Total impact on energy prices will, in the year 2000, amount to an increase of about 15 percent.[48]

Advanced Power Technology and Generation

The region lacks advanced technology for power generation efficiency and is falling increasingly behind the west. A striking case in point is aeroderivative gas turbine technology, the most-efficient and cost-effective solution for natural gas-fired power generation in the world today. Both Russia and Ukraine have advanced manufacturing capabilities for this system because it is a direct derivative of jet aircraft engine manufacturing. But U.S. and German manufacturers, General Electric (GE) and Siemens, for example, have much more efficient machines because western governments and manufacturers have invested more in research and development. An overhaul of GE's research and development management helped to move the "H-Class" turbine, which is intercooled, to market in half the time that it took for the "F-Class" turbine.[49] Western commercial turbine power generation systems reach efficiencies in excess of 50 percent, whereas eastern versions average 10–15 percentage points less.

Even before the collapse of the Soviet Union, Russian and U.S. scientists sought to create a state-of-the art gas turbine power generation industry.[50] The idea was to help create an alternative market for military technology, to reduce energy consumption, and to provide an alternative to nuclear power. The technology could hardly be more appropriate elsewhere, given Russia's possession of more than one-third of global reserves of natural gas. Russian experts suggested in the early 1990s that a $1 billion investment in the conversion of the existing Russian military jet engine industry to stationary power plant production could lead to production of 20 gigawatts of additional power capacity over the next few years. However attractive this investment may have looked on paper, it was always unlikely that U.S. or German firms would join in corporate ventures with these firms or even allow western assistance to permit improvements in eastern technologies. The former would require putting large sums of money at risk in unstable conditions. The latter would permit creation of formidable competition.

Two steps could help Russia and Ukraine better understand how to capture the potential of converting the military aircraft industry into producing clean-burning power plants. First, Russia would need to create the necessary institutional and legal infrastructure to provide incentives to generators to acquire highly-efficient power generation equipment. The best short-term alternative would be distributed aeroderivative gas turbines. Acquisition of U.S. experience in the Public Utilities Regulatory

Policy Act (PURPA) and in Integrated Resource Planning (IRP), including utility regulatory reform, would be useful. This expertise could be acquired on a commercial basis. Second, funding of aeroderivative gas turbine demonstration projects would be helpful. Although this expenditure might at first seem preposterous given the fiscal and financial crisis in Russia, Gazprom both has the capital and should have the motivation to participate.

Russian and Eastern European choices for power generation over the next decade include natural gas, coal, nuclear power, wind, and geothermal power systems (see Box 6.2). Solar power will mainly be useful off the established transmission grid. Gas has the disadvantage of higher fuel cost but the advantage of lower capital cost and much higher efficiencies, especially in combined cycle systems. Nuclear power enjoys very low fuel costs but high capital cost and a long construction period. Coal has both moderate fuel and capital costs, except when pollution control is included.

Typical costs of capital and fuel across the region can be combined with the expected efficiencies and operating costs of power generation systems to give a picture of the competitiveness of natural gas, coal, nuclear, and renewable energy (see Table 6.5). Natural gas combined cycle systems, at an efficiency of 45 percent and with a fuel cost of $65 per thousand cubic meters ($2.50 per gigajoule) and a capital cost of $750 per kilowatt, would provide power at a cost of $0.024 per kilowatt-hour. The actual cost will depend on the efficiency of power conversion, which Russian and Ukrainian turbines may not actually achieve.

Power from unwashed, unscrubbed coal-fired power plants would probably cost about $720 per kilowatt of capacity, and $30 per ton of coal. That would add up to about $0.03 per kilowatt-hour at the power plant boundary, assuming a power conversion efficiency of 37 percent. This way of stating cost excludes transmission and distribution costs, which vary across Russia, and is sometimes termed cost "at the busbar." Flue gas desulfurization would add about $150 per kilowatt and would increase the cost to $0.034 per kilowatt-hour.

New integrated, gasification of coal, and combined cycle systems (IGCC) cost at least one-third more than conventional coal-fired plants but are much more efficient. The fuel savings, however, do not offset the higher price of capital because IGCC power is estimated to cost $0.39 per kilowatt-hour. However, if sulfur removal is required—if the environmental externalities of sulfur dioxide emissions are included—the IGCC system is competitive with coal, although not with gas.

New Russian-designed nuclear power plant capacity costs about $1,450 per kilowatt of capacity. With low operating and minimal fuel costs, nuclear power would cost about $0.044 per kilowatt-hour. This cost estimate

---

BOX 6.2   A CASE STUDY OF WIND POWER IN UKRAINE

Ukraine sought to combine western renewable energy technology with its military's engineering and manufacturing skills. A joint venture called Windenergo Ltd., was formed in 1992 between Kenetech Windpower, Inc., and Pivdenmash, the Southern Mechanical Engineering Plant in Dnipropetrovsk, once the pride of the Soviet Union's missile industry. Wind turbines were built in a formerly high security shop where nuclear warheads for strategic missiles were assembled. The plant made a few 110 kilowatt machines and had plans to build 1,000 wind turbines per year.

Windenergo installed over 30 demonstration machines (3.6 megawatts) in May 1993 in the vicinity of Donuzlav Bay in Crimea, and the units proved reliable and efficient. Unfortunately, the demonstration was not well-sited, being placed more for the convenience of existing utilities at a naval operation than in the richest wind resource. Worse, Kenetech went bankrupt when the U.S. wind energy market collapsed after investment tax incentives for renewable energy were lost.

The windmill generators cost $720 per kilowatt, with operational cost at $70 per kilowatt annually. The annual capacity factor in the Crimean demonstration was about 25 percent. The resulting cost was $0.06–0.07 per kilowatt hour, with future costs expected to drop to $0.03–0.035 per kilowatt hour. Lack of financing and the non-payments problem, along with depressed demand for electricity, has dimmed the outlook for the joint venture.

SOURCE: Mykola Raptsun, personal communication, 1996.

---

assumes a nuclear plant is operated to western safety standards and takes about six years to build, less than the standard ten years in the west.

Transmission and distribution for all types of plants would add $0.01–0.02 per kilowatt-hour for industry and residences, respectively. In this calculation, operating costs have been added to all power systems based on experience in the region for each of them, and vary between $0.003 and $0.007 per kilowatt-hour.[51] Capacity factors are assumed to be 80 percent for gas; 75 percent for coal, IGCC, and nuclear; 50 percent for wind; and 80 percent for geothermal.

A promising new technology for the region is the fuel cell, especially given the abundance of natural gas available to the region. A fuel cell operates like an open battery. That is, instead of electrolytes generating current, fuel is applied to generate a current on electrodes separated by an

TABLE 6.5   Alternative Electric Power Supply Options

Power Source	Capital Cost (U.S. Dollars)	Lead Time (Years)	Cost[a] (U.S. Cents/kWh)
Gas Turbines	$750	2	2.4
Coal without scrubber	$725	4	3.0
Coal IGCC[b]	$1,025	4	3.9
Nuclear[c]	$1,450	6	4.4
Wind[d]	$750	2	4.0
Geothermal	$1,500	5	3.2

NOTES: a) Assumes 10 percent real discount rate, including construction period; lifetimes of 15, 30, 30, 15, and 30 years, respectively, for gas, coal, nuclear, wind, and geothermal systems. No transmission or distribution costs. b) IGCC = Integrated Gasification Combined Cycle. c) Decommissioning costs $300 million at closure discounted over plant life. d) 50 percent capacity factor, no storage.

SOURCE: William Chandler, Guo Yuan, Jeffrey Logan, Shi Yingyi, and Zhou Dadi, *China's Electric Power Options: An Analysis of Economic and Environmental Costs* (Pacific Northwest National Laboratory, Beijing Energy Efficiency Center, and China's Energy Research Institute, 1998).

electrolyte (see Figure 6.2.) For example, hydrogen can be produced from natural gas and used to generate current in a Proton Exchange Membrane (PEM) fuel cell. In the cell, hydrogen is ionized by a catalyst and brought in contact with an electrolyte, generating a current on the electrodes. Water is produced as a waste product, along with heat. The PEM system, which is undergoing rapid development, could be used in vehicle propulsion or stationary uses. A different but also promising new approach is the solid oxide fuel cell, which also produces waste heat at temperatures useful for space heating. Efficiencies for advanced systems range as high as 75 percent. Commercially available systems such as the Phosphoric Acid Fuel Cell, however, are 40–50 percent efficient. Research and development on these systems is underway in the United States, Japan, and, to a lesser extent, Russia. This system ranks among the most promising for reducing pollution and the cost of electric power production.[52]

Other sources, including hydroelectric power and biomass fuels, could be competitive in specialized circumstances, especially remote locations. But even considering the cost of transportation to Eastern Europe, natural gas appears to be the fuel of choice for the next decade, even if substantial fuel storage and costs are included.

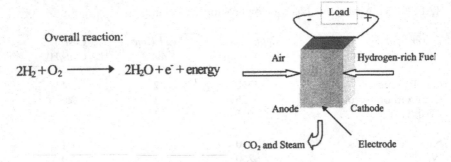

Overall reaction:

$$2H_2 + O_2 \longrightarrow 2H_2O + e^- + energy$$

Figure 6.2   Schematic Diagram of Fuel Cell Operation

Conclusions

Utility reform has achieved minimal success in the transition economies. Voucher privatization has been implemented in Russia and the Czech Republic, and Hungarian utilities have been sold off to western companies for cash. The concept of competition has been introduced, based largely on the British model. Competition works, more or less, at the wholesale level in Russia, Ukraine, Poland, and Hungary. However, neither regulatory agencies nor competition are providing accountability for electric utilities for most of the region.

World Bank analyst Dale Gray attributes this in part to a "lack of crisis . . . the lights aren't going out."[53] Gray also says utilities lack debt, which has been a motivating factor elsewhere for reform. Although it is not clear that private monopoly is better than a public one, it is also clear that the region can only regulate prices effectively in the short term. That is, regulating investment and retooling is beyond both the financial and technical capabilities of governments in the region. The privatization of distribution companies, typically in even worse technical and financial condition than the larger power generation companies, is probably years away. To use a medical metaphor, in the tradition of economists prescribing "shock therapy," emergency reform requires triage. Because there is only so much governments can do at one time, stopping the hemorrhage of money associated with the nonpayments problem is a priority. This effort will require additional support for structural adjustment, regulatory reform, tariff collection, accounting practices, and consumer assistance. Design of lifeline rates and direct assistance to residential customers will be required. Privatization and competition will likely be stalled for many years as a result of the Russian and Asian financial crises, which will make critically needed outside financing harder and more expensive to obtain.

Reform in the electric power sector requires less attention to specific technologies such as nuclear power and more attention—sustained attention—to systemic reform. As Russian electric power reform analyst Alexander Perevozchikov once said, "it is better to go in one direction slowly than to jump around in various directions." Three priorities stand out: resolving the excess power capacity problem because it interferes with efforts to create a functioning electricity wholesale market; making the wholesale system of tariffs and transactions transparent; and, most important, solving the nonpayments problem, which is necessary for fixing the others.[54]

Notes

1. One of the largest U.S. utilities, the Tennessee Valley Authority, is owned and controlled by the federal government. See, for example, William U. Chandler, *The Myth of TVA: Conservation and Development in the Tennessee Valley, 1933–83* (New York: Ballinger Publishing Company, 1984).

2. Neil Buckley and Simon Holberton, "Plug for the Generation Gap: Power Generators Fear the New EU Agreement on Energy Liberalization Unduly Favours the Monopolies," *Financial Times,* 26 June 1996, p. 13.

3. Emma Tucker, "Setback for EU as Plug Is Pulled on Current Project," *Financial Times,* 18 October 1996, p. 20.

4. The Ukrainian wholesale market in 1998 was stymied by widespread customer nonpayment for electricity and by serious fuel shortages that made supply uncertain. In 1999, more than one-third of all utility bills were unpaid, amounting to $1 billion in arrears, according to the Ukrainian State Statistics Bureau. "Ukrainians Delinquent on Utility Payments," *RFE/RL Newsline,* (4 February 1999); http://www.rferl.org/newsline/1999/02/040299.html.

5. Anonymous editorial, "Watchdogs Learn to Bite," *Financial Times,* 19 October 1996.

6. Simon Holberton and David Wighton, *Financial Times,* 4 December 1997, p. 7.

7. David Wighton, "Business Hits at New Limit on Gas-Fired Power," *Financial Times,* 11 May 1998, p. 7.

8. The United Kingdom early on grappled with the problem of anti-competitive practices in the utility supply market, particularly in dealing with long-term contracts between two major power generators, National Power and PowerGen, Eastern Electricity—with 6 gigawatts of nuclear plants—and fourteen power distribution companies. The fear was that unless distribution companies signed long-term contracts, they would lack secure and affordable power supplies, and that the existence of long-term contracts would prevent the entry into the market of new sources of power generation. Simon Holberton, "Electricity Competition Warning," *Financial Times,* 8 June 1997, p. 8.

9. Competition marks a shift from the integrated resources planning (IRP) model that was used by leading regions in the 1980s. This model differed substantially because it was developed for vertically integrated, privately owned utilities

with monopoly service rights for a specified region. It consisted of four basic steps. First, a demand projection was made using a process for public comment and revision. Second, a least-cost demand strategy was developed to allow for demand to be reduced. Third, a least-cost supply strategy was developed to satisfy unavoidable demand. Fourth, reconciliation of supply and demand was made, often by providing an environmental premium for demand-side management. That premium usually involved providing the utility company with a higher rate of return on investment in projects for promoting energy efficiency among its customers than was allowed for investment in power generating systems. The approach worked well by environmentalists' and residential consumers' estimation, resulting in reducing tens of billions of watts of power plant capacity requirements in the service areas of Pacific Gas & Electric, Southern California Edison, the Northwest Power Planning Council, and the New England Electric Company. Large consumers, utilities, and economists, however, would argue that enormous economic inefficiencies remain in this system. Both perspectives have merit, and finding a compromise between them—a system of competition that protects small consumers and the environment, while retaining efficiency—will be one of the great energy policy struggles of the next two or three decades.

10. *The Energy Newsbrief* 10, no. 2 (6 May 1996): 3.

11. World Bank, *From Plan to Market: World Development Report 1996* (New York: Oxford University Press, 1996.)

12. The earliest concerted effort was led by Marc Ledbetter of the American Council for an Energy-Efficient Economy and the Polish Foundation for Energy Efficiency (FEWE). With funding from the U.S.-based C.S. Mott Foundation, the groups organized a series of workshops in the region to introduce the concept of Integrated Resource Planning (IRP) to utility and distribution company officials, bringing U.S. experts for two weeks and translating a handbook on the design and implementation of IRP. As a result, the Polish government in October 1991 issued a statement endorsing the concept of IRP. A counterproposal would soon made by the British Know-How Fund. Working with the World Bank, the project sought to restructure the Polish electric utility industry. Poland tentatively adopted the British competitive utility model.

13. International Energy Agency, *Energy Policies of Poland* (Paris: Organisation for Economic Cooperation and Development, 1995), 115–117.

14. The International Finance Corporation (IFC) and the Global Environment Facility (GEF) demonstrated in Poland that distribution companies can improve their profitability by providing incentives for residential efficiency measures. The demonstration provided discounts to buy compact fluorescent bulbs, which are manufactured in Poland. Participation in the well-publicized program was high. Detailed end-use metering provided statistically valid data demonstrating reductions in peak-load demand for utilities, which reduced utilities' costs at the margin. These savings proved that utility incentive programs could increase utility profits without affecting power supply quality or creating burdensome administrative costs. Marc Ledbetter, Pacific Northwest National Laboratory, personal communication, Portland, Oregon, September 1998.

15. Polish Foundation for Energy Efficiency, *Poland's Alternative Energy Policy Until the Year 2020* (Warsaw, Katowice, Krakow, 1996).

16. International Energy Agency, *Energy Policies of Hungary: 1995 Survey* (Paris: Organization for Economic Cooperation and Development, 1995), 19–20.

17. John A. Riggs, Rapporteur, *Utility Restructuring in Central and Eastern Europe* (Washington, D.C.: Aspen Institute, 1997) (report of the 2d Central and Eastern European Energy Workshop, Klaus Brendow and Adam Guła, co-chairs, 7–10 May 1997, Krakow and Opuszna, Poland).

18. Virginia Marsh, "Hungary Delays Power Sale," *Financial Times*, 1 October 1996, p. 4.

19. Other insights are possible from plant capacity factors for the region, particularly with regard to operating older, dangerous, Soviet-designed nuclear power plants. In Ukraine and Russia, shutting down all the RBMKs and older VVERs would raise the capacity factor by only 1 and 4 percentage points, respectively, to 43 and 48 percent.

20. *RFE/RL Newsline*, 1, no. 21, pt. II (29 April 1997); http://www.rferl.org/newsline/1997/04/290497.html.

21. See SEVEn at www.svn.cz/bulletin-tables.htm, and Jirka Zeman, personal communication, Prague, Czech Republic, October 1999.

22. SEVEn, at www.svn.cz/bulletin-tables.htm.

23. Robert Anderson, "Czechs Hasten Energy Sell-off," *Financial Times*, 13 May 1999, p. 3.

24. Chrystia Freeland, "Kiev Power Reform Switched Back On," *Financial Times*, 6 June 1997, p. 7.

25. U.S. Embassy Kyiv, "Ukrainian Tender for Privatization of Power Generation Joint Stock Companies," *Bisnis* (Business Information Service for the Newly Independent States), U.S. Department of Commerce, 31 July 1997.

26. International Monetary Fund, Winter 1997, at www.imf.org.

27. Program Secretariat, Program to Promote Sustainable Development in Ukraine, U.S. Agency for International Development, Kyiv, "Tariff Policy Paper" (memorandum to Members of Energy Efficiency and Energy Saving Work Group, 5 May 1998).

28. Vitaly Sych, "Ukraine Stalls Energy-Sector Sell-Off," *Kiev Post*, 8 April 1999, http://www.thepost.kiev.ua/archive/ar05-os014-08apr1999.txt.

29. Cabinet of Ministers of Ukraine, "On Approving Financial Recovery Plan for the Power Sector of Ukraine," Resolution No. 508 (18 April 1998).

30. Citing *Stolichnye Novosti*, in *Bisnis*, 1 April 1998.

31. Russian Academy of Sciences, Energy Research Institute, U.S. Agency for International Development, *Joint Energy Alternatives Study*, draft final report, part II, U.S. Department of Energy (9 January 1995).

32. According to an analysis by Inne Gritcevitch, Program Manager with the Center for Energy Efficiency (CENEf), personal communication, Moscow, November 1999.

33. For comparison, note that the United States has 770 gigawatts of installed power generating capacity, of which nuclear and hydro amount to 100 gigawatts each. The majority—but not all—of the remainder is coal-fired. Western Europe, which has about the same population as the sum of the twenty-six transition economies, has about 600 gigawatts, of which 144 is hydroelectric and 123 is nuclear. U.S. Department of Energy, Energy Information Administration, June 1998.

34. Hagler-Bailly Consulting, *Joint Alternative Energy Strategies* (unpublished report 1996), 3–33.

35. Government Decree #1231 of 1997 states that reform in the housing and communal sector should be completed by 2003, including elimination of cross-subsidies in all utility services. Alexander Olegovich Perevozchikov, Senior Advisor, IRIS, Program on Natural Monopolies, personal communication, Moscow, November 1998.

36. Michael Specter, "Occupation of a Nuclear Plant Signals Russian Labor's Anger," 7 December 1996, pp. 1 and 6; Victor Yasmann, "Sit-in Strike at Russian Nuclear Power Stations," *RFE/RL Newsline* (8 August 1994), http://www.friends-partners.org/friends/news/omri/1994/08/940808.html; Penny Morvant, "Nuclear Workers Strike," *RFE/RL Newsline* (22 October 1996), http://www.friends-partners.org/friends/news/omri/1996/10/961022I.html.

37. IEA, *Energy Policies of Russia,* 202.

38. Chrystia Freeland, "Russia Power Market Launch," *Financial Times,* 30 June 1997, p. 2.

39. Chrystia Freeland, "Mosenergo Forced to Give Way," *Financial Times,* 18 April 1997, p.2.

40. Chrystia Freeland, "Soviet-era Power Bosses Sacked in Russia," *Financial Times,* 11 July 1997, p. 3.

41. Celestine Bohlen, "Energy Companies Are Tools in Russia Power Struggle," *New York Times,* 11 October 1999, p. A.3.

42. Cases of Russian utilities cutting off power to high-profile installations, such as a military radar in Khabarovsk in 1999, provide more theater than results for reform. "Local Electricity Supplier Pulls Plug on Military Radar," *RFE/RL Newsline* 3, no. 139, pt. I (20 July 1999); http://www.rferl.org/newsline/1999/07/100799.html.

43. Alexander Olegovich Perevozchikov, Senior Advisor, IRIS Program on Natural Monopolies, personal communication, Moscow, October and November 1998.

44. *OMRI Daily Digest* I, no. 1 (2 January 1997); http://www.friends-partners.org/friends/news/omri/1997/01/970102I.html.

45. Igor Bashmakov, "Heat, Money, and Housing Reform, *Energy Efficiency* 14 (January-March 1997): 1.

46. MFK Renaissance, *From the Sublime to the Ridiculous—A Credit Rating Model of Russia's Regions* (Moscow, April 1998).

47. Aleksandr Avdiushin et al., *No-Regrets Options in Climate Change Mitigation Policy: Lessons from Transition Economies,* Battelle, Pacific Northwest National Laboratory (May 1997).

48. *News at SEVEn* 3, no. 3 (June 1995): 1.

49. Stefan Wagstyl, "Blade Runner: The Fruits of GE's Research and Development Overhaul," *Financial Times,* 23 July 1996, p. 8.

50. Robert Socolow and Oleg Favorsky, "Aeroderivative Gas Turbines in the Soviet Union" (unpublished paper, January 1988).

51. Jeffrey Logan, Research Scientist, Pacific Northwest National Laboratory, personal communication (Washington, D.C.,1997–1998).

52. Sally Benson, William Chandler, Jae Edmonds, and Mark Levine, *Carbon Management: Fundamental Research Needs Assessment*, U.S. Department of Energy, Office of Energy Research (Washington, D.C., 1998).

53. Lights were "going out" in Ukraine, Georgia, and parts of Russia at this writing, however, due to the nonpayments and related crises.

54. These priorities were suggested by Alexander Olegovich Perevozchikov, Senior Advisor, IRIS, Program on Natural Monopolies, personal communication, Moscow, October 1998.

7

Environmental and Climatic Change

Jack Gibbons, physicist and presidential science adviser, once said, "There is no fundamental dichotomy between conservation and progress."[1] This comment challenges conventional wisdom that poor people cannot afford clean air and water. Sustainable economic development, if it means anything, achieves paired goals of economic development and environmental protection. But can climate change, which would occur over several decades, take priority when Siberia cannot afford heat, Ukraine refuses to shut down Chernobyl, and Poland lacks the investment to control acid rain? Ironically, the former Soviet bloc is well-placed to benefit from climate mitigation strategies that would help solve many other problems. Moreover, the transition economies may be especially vulnerable to climate change. Economic recovery and climate protection may be more closely coupled in the transition economies than is commonly supposed.

Six of the former Soviet and all of the Eastern European nations have signed and ratified the Framework Convention on Climate Change. They are Annex I countries pledged to early action.[2] The Kyoto Protocol of the Framework Convention on Climate Change (FCCC) further commits industrial nations to reduce their 1990-level greenhouse gas emissions by an average of 6 percent by 2008–2012. This agreement, negotiated by more than 150 participating nations, represents a significant step toward reducing the risk of climate change. Goals set at Kyoto vary by nation. Only industrialized countries, including the economies in transition, agreed to mandatory emissions reductions. The European Union (EU), the United States, and Japan agreed to cut emissions by 8, 7, and 6 percent, respectively, below 1990 levels between 2008 and 2012. Bulgaria, the Czech Republic, and Poland agreed to the EU target, but under the Framework Convention were permitted to choose 1988 as their base years. Emissions reductions would be measured against a "basket" of gases, including carbon dioxide, methane, nitrous oxide, hydrofluorocarbons, sulfur hexafluoride, and perfluorocarbons.[3] However, if Kyoto conferees agreed on a goal, they did not agree on how to reach that goal, and most countries have yet to ratify the protocol.

Certainly it matters that the region participate in climate policy, whatever form it eventually takes. Russia, Poland, and Ukraine possess one-quarter to one-third of the world's coal reserves, and coal produces more carbon dioxide per unit of fuel than any other. Russia and Central Asia also own 40 percent of the world's proven natural gas reserves. Burning gas produces half as much carbon dioxide per unit of energy produced as does coal, making gas a useful transition fuel and placing gas-rich nations at an advantage.[4]

The region possesses another "advantage" with respect to climate policy: gross energy inefficiency owing to central planning. Because international climate negotiators have set emissions reduction using 1990,[5] when the region's emissions were highest, participating in emissions reduction is both relatively easy and potentially advantageous. The region's large, cost-effective potential for efficiency improvement means that it probably offers the world's lowest cost source of greenhouse gas emissions reduction. Between 1988 and 1998, regional carbon dioxide emissions fell from 21 percent of the global energy-related total to just 13 percent (see Table 7.1). Because the Kyoto Protocol permits emissions trading, the transition economies might actually be in a position to benefit from climate protection.

Pecuniary gain may not be the only reason for the region to cooperate in climate policy. Climate change will likely affect northern latitudes disproportionately to the tropics and mid-latitudes. Unfortunately, much learning is required for policy makers to understand climate change policy because physics, ecology, and economics overlap to create a world of uncertainty. The complicated models of climate and energy sometimes seem only to prove Lewis Carroll's admonition that, "Nothing is so useless as a model on a one-to-one scale" (see Box 7.1). The future climate may be just that, making Earth a laboratory in which, as climate economist Jae Edmonds warned, "We are going to run the climate experiment."[6] Disarray in transition economies and governments could mean that early action to mitigate climate change will fail.

To understand these issues, policy makers must understand some of the underlying physics of climate change, its potential impacts, the applicable policy tools—economic modeling, cost analysis—and the methods and institutions for implementing climate policies.[7]

TABLE 7.1   Energy-Related Carbon Emissions by Country, 1988 and 1998
(Million Tons of Carbon)

	1988	*1999*
Russian Federation	646	407
Poland	134	89
Ukraine	184	91
Kazakhstan	69	31
Czech Republic	53	29
Romania	55	23
Uzbekistan	33	34
Belarus	33	17
Hungary	21	15
Other Former Soviet Union	49	14
Bulgaria	23	12
Slovakia	15	11
Azerbaijan	15	8
Turkmenistan	11	10
REGIONAL TOTAL	1,340	814
U.S. Total	1,245	1,640
Global Total	6,137	6,370

SOURCE: Author's estimates from British Petroleum, *Statistical Review of World Energy, 2000*, applying carbon coefficients of 13.7, 19.6, and 24.9 kilograms of carbon per GJ for gas, oil, and coal.

NOTE: Does not include non-carbon dioxide greenhouse gases.

Elements of Climate Change

The fundamental physics of climate change is fairly straightforward, but the term "greenhouse effect" is an imprecise metaphor. Greenhouses, pleasantly familiar sights in northern European latitudes, warm air in a way different from atmospheric warming caused by trace gases. In a vegetable greenhouse, sunlight shines through a glass or plastic exterior, is absorbed by plants, rocks, soil—whatever is inside the greenhouse— and the light is re-emitted as heat, which warms the air molecules. The glass, or plastic, traps warm air. Sunlight similarly shines through the Earth's atmosphere and warms the surface, which in turn emits infrared radiation, but as the infrared radiation is radiated back towards outer

---

BOX 7.1   GLOBAL CLIMATE MODELS

Climate models, also known as general circulation models (GCMs), solve global energy balances of sunlight and heat radiation back to space. The fundamental variables are direct solar light (S), diffuse solar light (s), surface reflectivity to sunlight, heat radiated from the atmosphere and the environment and reabsorbed by the earth's surface (I↓), heat radiated to space (I↑), the rate of heat exchange between the surface and the lower atmosphere (H), the latent heat of vaporization (L), the rate of exchange of water vapor between the surface and the lower atmosphere (E), the rate of heat exchange between the surface and the subsurface (G), and heat convection from surface water (F). That is:

$$(S + s)(1-\alpha) + I\downarrow \simeq I\uparrow + LE + G + F$$

The connection between greenhouse emissions and climate can come at several points. The most important is the absorption of long wave, or heat, radiation from the ground by greenhouse gases. These gases re-emit heat in all directions, including toward the ground where it increases surface temperature. Climate feedbacks can include greater cloud formation, and thus affect the reflection of sunlight. Higher clouds have a warming effect (they are cold and absorb more heat than they reflect as light) and lower clouds have a cooling effect (they are warm and absorb less heat than they reflect as light).

Global circulation computer models today can handle these mechanisms with relatively high resolution. They can model in the vertical dimension at several points in each column of air, and they can provide detail on a scale of one degree or less. However, that means that one can still determine regional impacts on, say, the Caspian basin, which is about that size, only with uncertainty.

---

space, trace gases, including water vapor, carbon dioxide, methane, nitrous oxide, and the artificial chlorofluorocarbons, absorb the infrared radiation. They in turn re-emit radiation in all directions, including back toward the ground, which increases surface temperature. As a result, the Earth's atmosphere is 20° Celsius warmer than it would be without the warming effect of the "greenhouse" gases. This much is well-established science.[8]

Also well-established is the fact that human beings have increased atmospheric concentrations of certain greenhouse trace gases. The atmospheric concentration of carbon dioxide has increased from 275 ppm to 360 ppm since 1750. Evidence is strong that this increase was caused by fossil energy combustion and deforestation. Methane is a more powerful greenhouse gas than carbon dioxide but is less concentrated in the atmosphere. Its concentration, however, has doubled since the Industrial Revolution, and again this increase is anthropogenic. The concentration of nitrous oxide, for reasons not entirely clear, has increased by 15 percent. In terms of radiative forcing, which is to say the overall increase in the greenhouse effect, carbon dioxide accounts for 65 percent, and methane and nitrous oxide account for 20 and 5 percent, respectively.

Atmospheric concentrations of the greenhouse gases are higher than they have ever been in the 160,000 years across which climate records exist. Evidence of climate change throughout geological time—the "climate record"—can be seen in indicators such as tree rings, sedimentary layering, and gases trapped in ice. Ice accumulates year upon year in Arctic and Antarctic fields, and the gases trapped in the ice vary in concentration as a function of temperature. The ratio of isotopes of oxygen (^{18}O to ^{16}O) provides a proxy record of temperature. Measurements can reveal evidence of volcanic activity—particulates and sulfur dioxide—as well as carbon dioxide concentrations. By drilling to great depths in the ice— more than 2 kilometers (1.2 miles)—climate records can be built that extend back hundreds of thousands of years. Correlations can be drawn between carbon dioxide concentration and temperature.[9]

The global climate during all of civilization has been relatively stable. Average temperature has not changed much—about 1° Celsius—over the past 10,000 years, which is about the age of Jericho, the world's oldest continuously occupied city. The Earth's average temperature has probably increased by 0.3–0.6° Celsius over the last 100 years. Global mean sea level at the same time rose 250 millimeters (almost 10 inches).

The current course of human events would now seem to be leading mechanistically toward a doubling—or more—of preindustrial greenhouse gas levels, the consequence of which will be climate change of significant and perhaps profound dimensions.[10] Controversy remains, however, about whether science is good enough to detect a "climate signal," an indisputable sign that climate change is already underway. Any signal that warming is underway should be quite small compared to natural variations in climate. In contrast, global average surface temperatures are projected by numerical models to increase by 1° to 3.5° Celsius (1.8–6.3° Fahrenheit) over the next century.[11] This rate of change would be 10–100 times greater than in the past 20,000 years. Arctic average temperature will increase faster than that of the globe as a whole. In Eurasia, winters appear to be shorter than 100 years ago. The Caspian and Aral Seas have risen and fallen, respectively, but for reasons either unrelated to climate or not yet understood (see Box 7.2.)

BOX 7.2   THE GREAT SEA BASINS

The Caspian Sea is rising and the Aral Sea is falling, both catastrophically. The Caspian's rise may or may not be related to climate change. The Aral's drop is clearly due to irrigation. The Aral Sea–the "sea of islands" in a local language–is located in an unpropitious spot for a lake. It is in the middle of a desert region that receives only 100 millimeters (4 inches) of rain per year. Four-fifths of the lake's water comes from the Amu Darya and Syr Darya rivers. The Aral's water comes from mountains to the east, in part from the world's two largest non-polar glaciers on Mount Khan Tengri, home to Snow Leopards, and in part from the Tien Shan mountains along the Chinese border. These sources feed the Amu Darya and Syr Darya Rivers, for millennia the two main sources for the Aral. These rivers were profoundly diverted for irrigation in about 1960, when the Aral began falling. The canals lose half their water through leaky tiles. Half the remaining water is used to grow cotton, a fifth for cattle feed. Irrigation for cotton here dates back a century, but water previously was taken from smaller branches. Some 90 percent of the rivers' water is removed to irrigate 3 million hectares of cotton, primarily, in the Fergana Valley of Uzbekistan and the more-southerly deserts of Turkmenistan. The volume of water reaching the Aral has fallen from sixty cubic kilometers per year to five. Summer inflow shrank essentially to zero, and the lake in turn has shrunk to half its 1960 volume, a level that had been maintained for the previous 10 thousand years. The drop, which has increased saline and mineral levels, has devastated the Aral's sturgeon, carp and other fisheries. Former fishing villages now are tens of kilometers from the sea.

The Caspian Sea presents a surprisingly different picture. The lake is bigger than California and has been rising very rapidly. Its water level rose 2.4 meters (8 feet) between 1978 to 1995, spreading water inland by 15-20 kilometers (9-12 miles). The Caspian, unlike the Aral, has not had a history of stability, fluctuating as much as 7 meters (23 feet) over the past two millennia. The Caspian fell from 26 meters (86 feet) below sea level in the early twentieth century to 29 meters (96 feet) below sea level in 1977. While underlying causes are not understood, the recent rise has been attributed to increased inflow and reduced evaporation.

The Caspian was once thought to be drying up so dramatically that the Arctic rivers had to be diverted to save it. This was part of "Stalin's Great Plan for the Transformation of Nature." The slogan of the region, according to conservationist Nicholas Denisov, changed from "Save the Caspian!" to "Save us from the Caspian!" Dried-out areas behind the retreating sea were developed without thought that the water level might rise again.

(continues)

(continued)

Three years after the sea level had started rising again, Soviet planners closed the strait between the Caspian and Kara-Bogaz-Gol Bay, but this allowed the Caspian to continue to rise while the outflow into the Kara-Bogaz-Gol was cut off—a safety valve closed. More significantly, the smaller body of water was cut off from the Caspian inflow and evaporated, harming the local chemical industry which used the salt water in production. Salt storms from the desiccated lake destroyed area top soil. The Turkmenistan government in 1992 ordered the dam removed.

But on the Russian shore of the Caspian, particularly in the western province of Dagestan, whole villages have been inundated. Rising water is expected to flood 100 more. The Volga delta is retreating north, and one of the world's largest wetlands is being submerged. The ecosystem cannot simply move north with the flood because the upper delta has been developed for agriculture and housing. Hundreds of oil pipelines and shut-in oil wells would be submerged with salt water. These facilities, not built to withstand corrosion, could create numerous, continuous oil leaks and severe environmental contamination.

The Caspian supplies four-fifths of the world's sturgeon, but the fish is sensitive to variation in salt, and levels now range between 13-200 parts per thousand. Wetland breeding areas were dried out and lost, poaching is severe, and Volga and Ural river dams have destroyed spawning. Hatcheries now must be used to maintain artificial reproduction. The Krasnovodski Zapovednik is said to have been overrun by hunters, with enforcement abdicated by the authorities, and the preserve abandoned to economic chaos. The chief Russian prosecutor tried to counter this trend in 1998 by blocking the Russian firm Lukoil from developing oil in the northern Caspian. His ruling was justified by the need to protect wildlife. It may have been coincidence that Lukoil had won the right to develop the field in a bitter fight with another oil company, Yukos, which filed a brief in favor of the fish.

The Euro-Siberian Climate

The climate in Eastern Europe and northern Eurasia is "profoundly continental," which means that weather patterns are deeply affected by the great land mass stretching between the Atlantic on the west and the Pacific on the east. Inner continental areas are marked by temperature extremes, and the lowest temperatures ever recorded in the northern hemisphere were registered in northeastern Siberia. The largest seasonal temperature swings anywhere occur in Siberia, with summer and winter average daily temperatures differing by more than 60° Celsius (108° Fahrenheit). In contrast, the range in Minnesota is more like 30° Celsius. Land cools more than oceans; that is, oceans can better store heat, and so the vast land mass of Euro-Siberia cools extensively and creates strong high pressure and cold, dry air.[12]

The number of frost-free days varies dramatically by latitude, averaging 200 per year on the Black Sea to just 75 in northern European Russia. Rainfall declines eastward from the Atlantic coast of Europe but does not differ dramatically from Warsaw and Moscow, which average 475 and 550 millimeters (19–22 inches) annually. In Siberia and Central Asia, annual rainfall averages as little as 250 millimeters (10 inches).[13] That is ironic because Western Siberia is the world's largest wetland, created by the late-season thaws and consequent flooding of the lower, more-northerly reaches of the great Siberian rivers running through Russia to the Arctic Sea.

Northern latitude temperatures are expected to increase disproportionately with global warming. Heat currently is carried northward from the tropics in part in an atmospheric circulation known as the Hadley cell. A Hadley cell can be pictured as hot air rising over the tropics, spreading northward over cooler areas and settling as high pressure over the mid-latitudes. The "Bermuda high," for example, is formed as part of this mechanism. Heated tropical air in the summertime rises from near the equator, sinks and warms over the area around Bermuda, forms a high pressure area rotating clockwise, and pumps insufferable summer humidity off the Gulf of Mexico and the Atlantic into Washington, D.C.

As in any heat engine, thermodynamics transfers heat from areas of higher to lower temperatures. Additional heat in the Earth's atmosphere would transfer heat disproportionately to cooler northern climes. This transfer would reduce the temperature difference between higher and lower latitudes, and because the jet stream is driven by the temperature differential between north and south—it is strongest, after all, during spring when that temperature difference is greatest—it would be diminished by this warming. Because the jet stream affects weather patterns and the distribution of precipitation, changing permanently the conditions that create the jet stream would almost certainly affect climate.

Warmer temperatures could also reduce periods of snow cover, which would have a feedback effect of reducing reflected sunlight—or increasing absorbed sunlight—and further increasing surface temperatures. More extreme climate change scenarios can also be imagined, with temperature shifts causing catastrophic shifts in ocean currents or prompting massive release of methane from clathrates along the continental shelf or in the permafrost in Siberia. One can imagine the impact on Murmansk, where the port is nearly ice-free all year due to the Gulf Stream, if there were a profound shift in ocean circulation. However, one need not accept the riskier aspects of climate change to appreciate its potential significance for Eastern Europe, Russia, and Central Asia.

A warming of 1–2° Celsius is projected for Siberia in the 2020–2050 period. This is a higher rate of warming than expected for the globe as a whole. According to climate models, this change would most likely be accompanied by a reduction in moisture and in peat production and boreal forest cover. The changes conceivably could result in a rapid oxidation of carbon from northern soils, thus increasing global anthropogenic carbon emissions by one-third.[14]

A warmer Euro-Siberian climate would affect agriculture, tundra, rivers, wetlands, wilderness, and all the human activities interlaced with snow and sun and rain. Tundra might be the most significantly affected ecological category. According to the Intergovernmental Panel on Climate Change, tundra is expected to exhibit high sensitivity to warming, with impacts including changes in the amount of frozen soil, levels of nutrients available, decreases in soil moisture, and shifts in species.[15] The oil and gas industries might be affected by changes in the permafrost because exploration and development depend to some degree on the ground remaining frozen. Development of oil fields, after all, is conducted over specially built "ice roads." Permafrost has shifted with temperature over geological time, moving north or south with changes in temperature at a rate of 60 kilometers per 1° Celsius (1.8° Fahrenheit).[16] Much of the truly permanent permafrost would disappear in European Russia, including in Yamal, where discontinuous permafrost would come to dominate.[17]

A foretaste of what may be in store is the recent experience of Sakha, formerly Yakutia, when permafrost melted because of unusually high temperatures. On the permafrost, foundations must be set on piers for stability. When permafrost thaws, piers become unstable. As a result of high temperatures during summer 1997, a swimming pool and a school building collapsed in the city of Yakutsk.[18]

The Siberian wetlands could also be affected by higher northern latitude temperatures, in the form of earlier melting of rivers and less flooding of Western Siberia. An increase in temperature in tundra areas might mean a shift to forest cover, provided sufficient water were available. Any decrease in the already limited precipitation in the Euro-Siberian north possibly could lead to desertification.

The effect of climate variation on rivers could be significant. Over the last 100 years, warming has probably contributed to an annual delay of freezing by as much as three weeks on the Danube, Dnieper, Don, and Volga Rivers. Winter ice on the upper Volga, Oka, Don, and upper Ob now breaks up seven to ten days earlier than late last century, although farther east, the Lower Yenisey and Lena Rivers stay frozen longer. Climate modelers project that break-up of ice occurs on average about five days earlier for every 1° Celsius (1.8° Fahrenheit) of warming.[19] Little is known about changes to expect in Arctic sea ice.[20]

The effect on forests could be extensive. Less than 20 percent of the boreal forest existed at the end of the last ice age. Only about fifteen tree species subsist in Siberia, and these vie within narrow climatic ranges of temperature and precipitation.[21] Boreal forests would be affected by climate-induced changes in rainfall (which could rise or fall by 20 percent), insects, and fires.[22] Some studies project that increases in fires could affect 7–12 million hectares (17–30 million acres), or 30–50 percent of the region's forested area.[23] The net loss of boreal forests in Russia could amount to one-quarter of the current area, and the total amount of current forest area disrupted or displaced could reach two-thirds. The current forest would, in effect, move north over a period of decades (see Figure 7.1).

Climatic change would detrimentally affect goals set by Eastern European countries, including Ukraine, to increase forest cover. Forests have been reduced over time in Poland and Ukraine to only about 10 and 15 percent of national land area, respectively, and both nations intend to increase this ratio. The Czech Republic has set a goal of increasing forest cover to 30 percent of land area, an effort that would be affected by shifting precipitation.

Massive ecological change, of course, would not be new to the continent. Society's direct impact on nature already overwhelms many ecosystems and will continue to do so for the foreseeable future. Current human activities already have significant impact on biodiversity, perhaps even outstripping the threat of climate change. For example, the Siberian Tiger once roamed from Beijing to Lake Baikal, but it is difficult today to imagine a tiger roaming near either. Only 200–300 remain, living in Primorski and Khabarovski krai.[24]

Common sense might lead one to expect that a warmer northern climate would benefit some human activities, particularly agriculture, in the former Soviet Union and Central Europe. Frost is often the limiting factor in crop yield in temperate climates, which would include Central Europe and much of European Russia. Higher minimum temperatures in spring would likely boost yields. Moreover, carbon dioxide is itself a nutrient for plants, and increasing its concentration in air would have a fertilizing effect, improving yields in some grains by perhaps 30 percent, if carbon dioxide concentrations were increased to 575 ppm.[25] Farmers have flexibility in adapting to climate change because they already deal with weather variability.

However, a sanguine agricultural outlook is not supported by climate science. Higher maximum temperatures could adversely affect crops

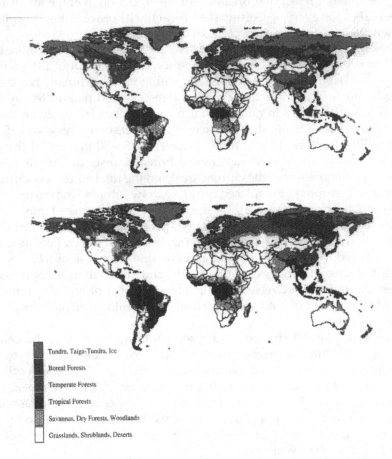

Figure 7.1   Potential Climate Change Impact.

NOTE: The top and bottom figures represent current and future distributions of biomes, respectively, where the future projection simulates possible results with a 575 ppm atmospheric carbon dioxide concentration. The preindustrial and current levels total 275 and 365 ppm, respectively.

SOURCE: Robert Watson, Marufu C. Zinyowera, and Richard H. Moss, eds., *Climate Change 1995: Impacts, Adaptations, and Mitigation of Climate Change: Scientific-Technical Analyses,* Contribution of Working Group II to the Second Assessment Report of the Intergovernmental Panel on Climate Change (Cambridge: Cambridge University Press, 1996).

such as potatoes by depressing growth, which is known to happen if average high temperatures range much above 20° Celsius.[26] Some vegetables such as spinach and broccoli prefer cool nights. Changes in temperatures and rainfall may affect soils and yields and may increase problems with pests. Crops obviously need moisture, but climate models are notorious for their inconsistency in projecting changes in rainfall patterns. The problem is that no one knows how precipitation will be affected by global warming, and precipitation is obviously a vital factor in agricultural productivity. Different models, such as those of the Princeton Geophysical Fluid Dynamics Laboratory (GFDL) or the United Kingdom Meteorological Office (UKMO), may show opposite trends in temperate and polar regional precipitation. Rain may fall harder and therefore be less available to crops. Yields in Central Asia, southern Western Siberia, and the North Caucasus also differ widely depending on the model used: +50 percent for GISS, −2 to +25 for GFDL.[27]

Yields for grasslands are generally expected to decline. Studies for Eastern Europe are rare.[28] Higher temperatures would lead to higher evapotransporation rates and therefore greater water demand. Weeds and pests would likely migrate north. Yet, the impact of water, soils, and pest management depends as much on humans as on the climate, and so the uncertainty in projections would remain even if the models were more consistent and had higher resolution, or ability to focus on smaller areas.

The current generation of climate models describes a wide—one could say wild—range of agricultural impacts for the former Soviet Union.[29] Wheat production in the year 2030 could be 19 percent lower or 41 percent higher, depending on the model. Similarly, corn production could be lower or higher.[30] Climate change, insofar as regional agriculture is concerned, would appear to be a high-risk game.

Energy Projections, Climate Forecasts

Economists sometimes protest that they do not forecast the future, but merely "project" it. Like weather forecasts longer than seventy-two hours, energy projections are uncertain. But as financial managers who invested in emerging markets could attest, it is useful to anticipate the future to minimize risk.

Three challenges confront those who grapple with the benefits and costs of climate change mitigation in the region. First, no one can estimate the benefits because no one knows the extent of damage—or conceivable benefits—that climate change would bring. Second, economic modeling suffers from well-known uncertainties about the future of technology, human behavior, and the effectiveness of policy. Third, the region is so profoundly distorted economically that the difficulty of modeling "normal" economies is greatly compounded. Who will predict the future of Belarus or Ukraine? Who knows whether Gazprom will sustain its monopoly, if Poland can bear to lay off a hundred thousand miners, or even—to pick a simple matter—whether the world oil price will rise?

A number of economists have tried. Analysts have produced a range of results, however, reminiscent of the forecast for the impact of climate on regional agriculture: It may be good, it may be bad. A methodological reason, however, can be found for the differences in energy and economic evaluations of climate policy (see Box 7.3). Top-down energy-economic models typically yield net economic costs, whereas bottom-up models yield net economic benefits. A similar outcome results from the choice of methodology for modeling other parts of the world. However, the nature of planned and post-planned economies increases the difference between the two results using the two approaches.

Reform and restructuring are handled differently in the two model types. The strength of macroeconomic models is that they can incorporate "feedback effects." They can model economic growth, fuel price changes, and changing energy demand in an internally consistent manner and in a way consistent with economic behavioral theory. Growth and prices generated by the models are endogenous results, whereas bottom-up models usually assume the values of these variables. The bottom-up models, like the name suggests, handle much more detail about energy supply and demand systems, particularly with regard to technology. Bottom-up models can estimate energy demand on the basis of economic growth, structural change, price response, and technical energy efficiency improvements not attributed to price response. These models typically are driven by assumed rates of economic growth and, like macroeconomic models, also have exogenously assumed rates of technical change.

Two examples illustrate the more general shortcomings of economic modeling, but especially as it applies to the transition economies. In the first example, automobile efficiency is improved by fiat, and energy demand is reduced relative to the case in which efficiency does not improve. But reduced demand would also in theory reduce gasoline prices, reducing consumers' concern for saving energy and offsetting some of the savings produced by the fuel economy standard.

Incorporating such feedback effects is important to achieving desired policy results and for making energy supply—and demand—investments. The fuel economy standards imposed in the United States in 1975 doubled the fuel economy of passenger cars. However, as oil prices sank, consumers moved to less-efficient trucks. Light trucks were 50 percent more efficient than old cars, but they were one-fifth less efficient than

---

BOX 7.3   ELEMENTS OF ENERGY-ECONOMIC MODELS

To be valid, economic models must be based on sound economic principles, valid assumptions, and they must be transparent. A key question is "What drives the model?" What are the functions that create change, and what are their variables? Extrapolation of past trends and use of parameters unrelated to behavior are often not credible.

Most macroeconomic energy models will have GDP growth as a driver, with GDP itself driven by variables such as growth in the labor force and growth in productivity. For example, GDP can be defined as the number of workers (W) times labor productivity per worker (LP), or:

$$GDP = W * LP$$

Energy demand or changes in energy demand can next be modeled as a function of key variables including GDP per capita (GDP/Population = X), the income elasticity of energy demand (y), energy price (P), energy price elasticity of demand (p), and population (Pop). Recall that price elasticity is the percentage change in energy demand for every 1 percent change in energy price, and that income elasticity of demand is the percentage change in energy demand for every 1 percent change in GDP per capita. The price elasticity, for example, is usually calculated over a set of times series data as the correlation coefficient of energy demand on energy price (actually, the square of the correlation coefficient, or R^2). This parameter is a logarithm, and is written as an exponent. For example, economists have often found that correlations of demand and price changes show that demand for oil changes by 0.7 percent for every 1 percent change in energy price. If a price index for oil doubled–from 1 to 2–the change in the index of demand for oil (initially demand = Y = 1) would be 0.62:

$$Y = P^{-\epsilon} \qquad \{Y = 2^{-0.7} = 0.62\}$$

In words, demand falls from 100 percent to 62 percent.

A simple model for energy demand (Y) can be put together from changes in GDP (X), energy price (P), and the estimated GDP elasticity of demand (y), the estimated price elasticity of demand (p), and population (Pop), like this:

$$E = X^y * P^p * Pop$$

An equilibrium model would produce its own prices endogenously–internal to the model–rather than assume them exogenously–that is, take them from the literature. A model can still be

(continues)

(continued)

valid if key variables and parameters are assumed, as long as they are credible and, more importantly, as long as they are transparent.

A bottom-up model would assume these variables and combine them with assumptions for technical efficiency. Energy demand for the steel sector (E_{steel}), for example, could be modeled by assuming economic output (X); tons of steel produced per dollar of economic output; the rate of change s in steel production as a function of economic growth (to reflect structural change through reform); future energy price, P; the price elasticity of demand p; and the annual rate of reduction in the energy intensity (EI) of steel production (to reflect technical change), to give:

$$E_{steel} = X^s * P^p * EI$$

new cars. Consequently, the fuel economy of new U.S. light passenger vehicles improved from 14 miles per gallon to almost 28 miles per gallon, only to fall back to 21 or 22 miles per gallon as consumers substituted vans and trucks for cars. Macroeconomic models would not have been able to incorporate explicit fuel economy policy, but if they could, they would have predicted the fuel price softening. Bottom-up models could explicitly have managed the fuel economy policy, but usually cannot handle endogenously the behavioral impact of softening prices. Although this can be handled exogenously by assuming the price changes, the result is less than elegant.

In the second example, a Russian steel mill initially requires twice as much energy as a Japanese or U.S. one to make its product, but improves as it faces international competition arising with economic reform. The bottom-up modeler could incorporate a rate of improvement due to competition-induced modernization at a rate consistent with introduction of competition and with the steel sector's ability to finance the improvements. The improvement rates can be estimated in detailed sectoral and market studies. Top-down, or macroeconomic, modelers would assume that the Soviet economy was already in competitive equilibrium: that the steel mill was already economically competitive. There is an international market, after all, for relatively low quality but inexpensive steel. But starting from an economic baseline of irrationally high energy intensity, macroeconomic modelers often incorporate no reform or restructuring. Many of the best-known energy and climate studies have treated the formerly planned economies in exactly this fashion.[31] Because this practice is not consistent with reality or even economic theory and is impossible to accept empirically, the result too would be unacceptable. The modeler, of course, could modify baseline energy demand arbitrarily to anticipate economic reform, but this, too, would be inelegant. Indeed, the model would no longer be a proper general equilibrium model, but a bottom-up model, most likely lacking transparency in handling reform.

Early political perceptions of the cost of climate policy were colored by macroeconomic modelers. Famous studies led by U.S. economists Alan Manne and Rich Richels treated the former Soviet Union as a separate region in the "Global 2100" model and estimated the costs of reducing energy-related carbon emissions.[32] That top-down modeling produced a set of regional results for a base case and for significant emissions reductions relative to the base line. The results suggested that the cost of emissions reduction would total 3 percent of regional Gross National Product (GNP) in the first half of the next century. Other, similar approaches reached similar conclusions.

Subsequent work suggested that that cost estimate was too high. Several "bottom-up" studies have been completed by Eastern European and Russian experts since the revolutions of 1989. An effort led by Russian academician and economist Alexei Makarov used a dynamic optimization model developed specifically for the former Soviet Union. A projection to the year 2025 suggested that a 44 percent emissions reduction, compared to a baseline, would cost less than 1 percent of GNP.[33] Like Makarov's team, and unlike the top-down studies mentioned above, most of the bottom-up studies incorporated economic restructuring. Clearly what is needed is a model that combines the best of both approaches. Some laboratories have been developing such models, but little serious attention has been devoted to this problem.[34]

Both types of models must solve the problems of consumer behavior. How much will consumers cut energy demand as prices increase? How much will they increase demand as their incomes rise? Energy price elasticities in most models must be assumed exogenously using empirical studies. One celebrated case involved a consultant for the European Bank for Reconstruction and Development on the need for power to replace Chernobyl, who addressed this question·by performing a regression analysis of energy prices in Ukraine over the 1970s and 1980s. The result was an estimate that the elasticity was nearly nothing, that no matter what the price increase, Ukrainians would continue to use energy at the same rate because their economic situation made them unresponsive to prices. Of course, prices in Ukraine were meaningless throughout that period. They did not reflect marginal costs, fuel was rationed, and payments were simply a means of keeping accounts. The prices, in fact, changed only twice over the period, and then only slightly. Money not spent had little value—there was no opportunity cost because investment was impossible and saving more or less was meaningless—further diminishing the meaning of price response. The consultant's elasticity was simply a measure of consumer response to meaningless prices for a rationed, allocated good. A better, if imperfect, approach is to assume that Ukrainians and Russians and Poles will react to prices the way consumers do everywhere: When prices go up, demand goes down. In Western Europe, Japan, and the United States the ratio, or price elasticity of demand, seems to hover around –0.8.[35]

Estimating income elasticity of demand is also difficult because there are countervailing pressures in the region's economies. As state-owned enterprises are privatized or closed, and as higher value-added light manufacturing and services are added, less energy-intensive activities will replace them. Technologies that use twice as much energy per ton of steel or block of cement will be replaced by technologies similar to those in the west, or producers will be unable to compete with international firms. Probably it would be safer to project a range of futures using income elasticities of 0.5 to 1.0. That is, one could assume first that total energy demand will increase only half as fast as income, and then assume that both increase at the same rate. Bottom-up modeling provides a more detailed analysis of policy and technical options than can top-down modeling, but it does a poorer job of accounting for economic behavior. Unfortunately, some bottom-up models do not sufficiently incorporate a "theory of change," that is, an explicit behavioral mechanism for market penetration of emissions-reducing technologies.

The word *mitigation* is used to mean preventing climate change, as opposed to *adaptation*, which means offsetting its effects. The importance of mitigation for the formerly planned economies is their great wealth of energy-savings opportunities. Alternative energy options, except for low-carbon natural gas, simply cost too much at present, and the demand for them, except for off-grid applications, is too depressed. Available mitigation options in the region include the following:

- Energy efficiency
- Switching to natural gas
- Nuclear power
- Renewable energy
- Afforestation

Regional experts have evaluated these measures in systematic studies and in extensive detail (see Table 7.2). Most of this work involved a cross-section of economists, technical experts, policy analysts, government representatives, and environmentalists to give the studies the perspectives not only of academicians but also of stakeholders. The results in general

TABLE 7.2   Estimated Cost of Carbon Emissions Reduction in the Polish
Economy

Measure	($/Ton C)
Industrial Efficiency	
Workplace lighting efficiency	-36.90
Automated process controls	-36.00
Electric motor efficiency	-34.00
Steam management	-33.60
Electronic energy controls	-32.40
Blast furnace gas turbine	-31.20
Energy management	-30.80
Steel plant modernization	-24.80
Heat recovery	-20.00
Furnace modernization	-19.20
Waste heat recovery	-18.00
Dry process cement production	-17.20
Buildings/Urban Efficiency	
Street lighting efficiency	-66.00
Controls for gas boilers	-60.80
Commercial lighting	-58.40
Home lighting efficiency	-55.60
Household appliances	-40.80
Home wall insulation	-34.40
Window weatherization	-34.40
Public building insulation	-29.60
Roof insulation	-24.40
Renewable Energy	
Home heating, fuelwood	-35.60
Home heating, crop residues	-34.40
Biofuels, rapeseed	-19.20
Geothermal heat and power	-10.40

Source: Adapted by the author from Janusz Michalik, Sławomir Pasierb, Jerzy Piszczek,
Michał Pyka, and Jan Surówka, *Evaluation of the Feasibility and Profitability of Implementing
New Energy Conservation Technologies in Poland* (Polish Foundation for Energy Efficiency
[FEWE], Katowice, Poland: 4 May 1994); assumes coal as the energy source.

are strikingly similar, although this is not surprising given the uniformity of Stalinist economies. The factories, the buildings, and the cars all had a uniform technological level.

A study in Poland led by economist Henryk Gaj[36] provides a typical set of measures, along with their costs and effectiveness. These were ranked by least cost, with costs given mainly in negative U.S. dollars, which means they reflect not costs, but cost savings. Poland's energy intensity has fallen by an estimated 25 percent since 1989 as it has implemented restructuring, price rationalization, and reform, giving credibility to the feasibility of implementing such measures.

For example, improving street lighting saves electricity and the carbon produced when coal is burned to generate the electricity. Because switching from mercury vapor to sodium vapor lamps can save half the electricity required to produce power, half the carbon normally emitted when coal is burned to illuminate streets can be saved. That translates into a cost savings of $66 per ton of carbon saved, even accounting for differences in the costs of the lighting fixtures.

There are two distinct but confusing aspects to discount rates for energy alternatives, and this distinction often lies at the heart of controversy between top-down and bottom-up modelers. One key issue relates to the concept of consumer surplus. Some top-down modelers argue that markets are in competitive equilibrium and that any deflection from that path by definition imposes economic costs. This assumption fails the test of economic validity both on theoretical and empirical grounds.

Bottom-up and top-down modelers would probably agree that the apparent consumer discount rate for energy efficiency is quite high—over 50 percent in many sectors—meaning that there is a high preference for consuming now rather than in the future. The controversy arises in the interpretation of the meaning of this discount rate. Some economists argue that the apparent rate reflects real preferences among consumers, and that historic energy use patterns in market economies are optimized for public welfare. Other economists would argue, however, that the apparent rates reflect only distortions in the economy and can be greatly reduced. This latter group points to subsidies for energy use, regulation of electric utility prices, and lack of consumer financing as examples. The discount rate in their work is used only to incorporate the time value of money and may be as low as 3 percent to reflect the long-term cost of capital.

A Ukrainian team detailed similar mitigation options in Ukraine.[37] The team estimated emissions reduction potential of 100 million tons of carbon per year in 2015. The study assumes full economic recovery and robust growth by that time. Ukraine is a significant player in global climate change, ranking in the top dozen nations in energy-related greenhouse gas emissions.

A case study from Russia suggested that Russia could reduce energy intensity with reform.[38] The study was of a private Russian manufacturer, Moscabelmet, which employs 1,500 in the manufacture of aluminum and copper cables and wires. It is one of the nation's largest producers of these products. The economic reforms have forced up the share of energy costs as a percentage of total production costs from less than 5 percent in 1993 to 15 percent or more in 1995. The company specifically set out to save energy to save money. It took the following steps:

1. Conducted an energy audit;
2. Changed operations using an energy monitoring system;
3. Installed steam traps, heat recuperators, and pipe insulation; and
4. Introduced automated energy control.

Despite the general turmoil in Russia, the firm remained current with tax payments and energy and utility bills. Production levels were maintained, wages were paid on time, and layoffs were avoided over a two-year period. The automated control measures and operational changes are applicable in industry throughout the region.

A Russian climate mitigation study team also found much potential for carbon emissions reduction in the energy supply sector (see Table 7.3). A variety of measures such as upgrading the compressors on gas transmission pipelines, installing better turbines for power generation, reducing losses in district heating, and installing catalytic cracking and better distillation columns in refineries would save almost 100 million tons of carbon per year. That is equal to one-fifth of Russia's total emissions in the late 1990s.

A similar team in the Czech Republic identified almost a score of priority measures for cheaply cutting carbon dioxide emissions.[39] The items fell into three broad categories: incentives and disincentives for energy use, financial assistance for investment, and standards and information. The three most effective items involved eliminating preferential tax treatment for energy, providing incentives for reforestation, and facilitating leasing of energy efficiency equipment. Also recommended was shifting district heating subsidies to incentives for energy efficiency. The Czech government provided a $3 per gigajoule subsidy for residential heating until 1998, when all heat subsidies were eliminated. Czech industrial energy customers paid about $9.40 for the same quantity of heat.

Energy Sector Considerations

Energy and climate policy modeling involves the problem of fuel choice. Coal, oil, and natural gas and their products release different amounts of

TABLE 7.3   Cost-Effective Carbon Emissions Mitigation Opportunities in the Russian Energy Sector

Sector	Savings Potential (EJ)	Emissions Reduction (million tons, C)
Power generation	2.6	50
District heating	1.2	24
Oil wells, refineries, pipelines	0.8	18
Gas wells, pipelines	0.7	15
Coal bed methane	0.1	3
Total	5.4	110

SOURCE: "Energy Strategy of Russian Federation" (Ministry of Fuels and Energy, Moscow, 1995, mimeographed, translation provided by Tatiana Shiskina, Center for Energy Efficiency [CENEf], Moscow, January 1996).

carbon per unit of energy produced. Natural gas produces only 55 percent as much carbon dioxide for a given unit of energy as coal does (see Table 7.4). Natural gas can often be used more efficiently than coal for the same applications. Incorporating efficiency of combustion means that gas generally produces less than half as much carbon per unit of energy used.

Total projected emissions of carbon dioxide to the atmosphere from energy use are simply the total of the fossil energy demand projected by type multiplied by the carbon emissions coefficients for each fuel type. Energy-economic models can be linked to climatic models, or general circulation models, by adding the radiative absorption effect of emissions projections. Significantly, only about half of the annual anthropogenic addition of carbon remains in the atmosphere because plants and the oceans absorb and sequester an amount equal to half that total.

Scenario Analysis

Exploring the future with energy-economic models is best done in scenarios. This practice establishes a baseline projection, or scenario, against which little experiments can be run by testing the change in energy use, say, or carbon emissions, resulting from the changes in assumptions. The term *scenario*, borrowed from the theater, describes an effort to tell a consistent, transparent story about the future. For a baseline scenario, one would normally generate a "business as usual" case, which more or less

TABLE 7.4 Carbon Content of Fossil Fuels

Fuel	kg-C/GJ	lbs-C/MMBTU
Coal	24.9	54.8
Oil	19.6	43.1
Natural gas	13.7	30.1

KEY: kg = kilogram; lbs = pounds; C = Carbon; MMBTU = Million British Thermal Units.

NOTE: Units are for kilograms of carbon—as opposed to carbon dioxide—per gigajoule or pounds per million BTU. Carbon dioxide per unit of energy can be obtained by multiplying each carbon value by 3.67, the ratio of the molecular weight of CO_2 to C.

SOURCE: Gregg Marland, Carbon Dioxide Information Center, Oak Ridge National Laboratory, personal communications, Oak Ridge, Tennessee, June 1976, June 1988, and October 1990.

continues past trends. The IPCC in 1992 created a global baseline in which global energy use grew from a starting point of 385 EJ to 1,000 EJ by the year 2050, a growth rate of 1.6 percent.[40] Global carbon emissions doubled over the period from 6 to 12 billion tons. One can examine the assumptions for the key variables and make different assumptions that would achieve a more desirable output. The key to policy in this type of exercise, however, is to articulate a theory of change. One cannot just assume that technology will be used because it is available, and one must articulate why prices, technology, purchases, and preferences would change. The only tools available for such analysis—and the only tools available for governments to change the world—are research and development, information, standards, taxes, and incentives.

Creating a baseline for the post-planned economies from historic trends would be nearly impossible and essentially meaningless. Russian economist Igor Bashmakov has proposed a "business as unusual" approach, which tests reform in differing degrees. In other words, the modeler can craft scenarios ranging from "stagnation with little reform" to "dynamic growth with rapid reform." One can also have "emissions trading" scenarios in which the Kyoto Protocol on Climate Change is implemented. Other policies, ranging from carbon and energy taxes to efficiency standards, can also be tested.

The transition to markets, however, changes all the old assumptions, including energy-using behavior and the very structure of the region's economy. The IPCC scenario, which has been the basis for much of the international community's negotiations of emissions limits, pegged the region's carbon dioxide emissions at the end of the 1990s at about 1.5 billion tons. Actual emissions in 1999 were half that. Indeed, the region's share of global energy-related carbon emissions has fallen by almost 40 percent over the 1990s. The regional total, about 814 million tons of carbon out of a global total of 6.4 billion tons, compares favorably with the United States, which emits almost twice as much with roughly the same population.[41]

The transition economies must reform, restructure, and rationalize their economies if they are to grow.[42] If they succeed, energy intensity will dramatically fall. If they do not, growth will not increase emissions because growth will not occur. One IPCC future energy scenario for the region assumes continued high-energy intensity and high economic growth (see Figure 7.2). The results are startling when one realizes that they indicate that the formerly planned economies would be twice as energy intensive in the year 2050 as Japan is today. This outcome is highly unlikely without the restructuring and efficiency improvements necessary to enable formerly planned economies to compete internationally. Although one should not overstate the difficulty of reducing emissions in the transition economies, some of their achievements have been impressive. Not many policy makers—certainly not in the United States and probably not even in Europe—could change their energy prices from essentially zero to world levels in a few short years.

Reconsidering the region's climate future is important for global climate policy. Because climate literature forecasts—which did not foresee the collapse of the Soviet Union—overestimate emissions by almost 10 percent of global emissions, the baseline against which all near-term emissions reductions is measured is too high, meaning that cost estimates are too high as well. Planned economies are obviously far below the ceilings they agreed to when signing the Framework Convention on Climate change. These countries will have large quantities of low-cost emissions to trade, if trading mechanisms are established. Whether this situation is positive or negative depends on one's perspective. Europeans and some environmentalists consider this vacuous trading of "hot air" that would enable the region to sell emissions that they will not make in any event, and enable the west, principally the United States, to meet Kyoto goals by doing virtually nothing.

In this context, it is important not to lose sight of the longer term issues. It is by no means clear that the region will not return to 1990 emissions levels over the next two decades. General economic reform and recovery coupled with the failure to reform in the energy sector could make possible a more energy-intensive economic structure than would result if the energy sector were also reformed. Heavy industrial production could be restored, capturing markets at home and abroad for steel,

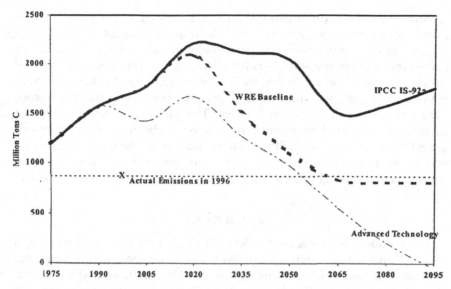

Figure 7.2   Carbon Scenarios for the Transition Economies

NOTE: WRE=Wigley, Richels, Edmonds; IPCC=Intergovernmental Panel on Climate Change.

SOURCE: Jae Edmonds and William Chandler, Pacific Northwest National Laboratory.

aluminum, chemicals, and cement. Household energy could easily surge as the number of cars and distances traveled increase with wealth, home sizes and appliance numbers grow, and air travel expands. The number of private cars in Eastern Europe has already increased dramatically.

On the other hand, reform coupled with climate cooperation with the developed world could effect emissions reductions relative to almost any plausible baseline.[43] What will it take in the way of investment, policy, and institution building to enable the transition economies simply to become at least as energy-efficient as the United States by the year 2020? Achieving this result would require massive increases in investment, although the net savings would total over $1.5 billion per year. The net savings over the period would be worth over $35 billion.[44]

After 2020, even this region would have exhausted the easy energy intensity reducing measures, and then new technologies must make a contribution. The lowest scenario tests a case in which fuel cells, powered by hydrogen derived from biomass, become cheap enough to compete with coal. All such scenarios have at base the assumption that the west will continue to invest in research and development and that technology transfer through the marketplace will take care of technology diffusion. This assumption has several flaws. First, western nations have been cutting their research budgets to reduce their government deficits. Second, technology development usually does not happen in a vacuum but requires some enabling atmosphere. Clean air standards in the United States, for example, prompted the development of affordable catalytic converters for cars and scrubbers for power plants. The Montreal Protocol prompted—and was facilitated by—development of substitutes for chlorinated fluorocarbons. Third, technology does not simply transfer in container ships carrying new technology. The production, utilization, and adaptation of technology requires people to have been trained and experienced in technology, and without an investment in the formerly planned economies in training and education commensurate with the new technology needs, the technology transfer needed is not likely to come about.

Beyond Kyoto

The key issues left undecided at Kyoto include emissions trading, use of "sinks," and enforcement. Emissions trading became a contentious and emotional issue, pitting Europe against the United States and the transition economies. Some European and developing country participants opposed trading because it created a potential loophole; others found trading immoral. The term "hot air" characterized this debate. The debate stemmed from fear that phantom emissions reductions—hot air—could be traded by Russia and others. By buying reductions made as a result of economic collapse since 1990, the United States, especially, could purchase a cheap way of meeting its targets without reducing its own emissions. The moral argument was that one ought to clean up one's own emissions, not hire someone else to do it. Supporters of emissions trading argue that it would reduce the cost of emissions reductions and provide badly needed financing for the transition economies.[45] Trading would, nevertheless, require safeguards, such as tying credits to specific investment projects, to ensure that real, additional emissions reductions are achieved.

The "sinks" debate was one of how to count deforestation, afforestation, and reforestation in a nation's progress toward meeting its emissions reduction goal.[46] This discussion also became emotional. Some delegates objected that science's ability to measure changes in sinks is inadequate to avoid abuse of this mechanism. Environmentalists objected that including sinks might give an incentive to countries to cut old-growth forests—which take up less carbon per year than young forests—and replace them with rapidly growing monocultures that do not support biological diversity.

Kyoto authorized emissions trading under the convention, but limited the practice to developed countries. Delegates postponed consideration of rules to govern trading. They agreed to incorporate changes in sinks in national emissions limits but restricted the definition of sinks to include only forests and deferred the controversial question of measuring changes in sinks. The representatives put off defining any enforcement mechanism for emissions limits, although they suggested these may include fines or payments to a "clean development fund." Unfinished Kyoto business will demand the attention of the transition economies for years to come.

Working out mechanisms for joint implementation is a priority. The term in the formal sense means that companies from foreign countries could share emissions reduction credits with local firms under a future greenhouse gas emissions convention. Governments like that of Ukraine have higher priorities for economic growth, and they may not have a clear picture of the benefits of and mechanisms for allocation of emissions reductions.[47]

Implementing mitigation measures through energy efficiency will require realistic energy pricing, hard budget constraints, financing, and technical expertise. Some progress has been made toward meeting these prerequisites. Russia, Ukraine, and Poland account for 72 percent of the greenhouse gas emissions of the former Soviet Union and Eastern Europe, and so are the focus of climate change policy in the region. The Czech Republic has been a leader in this area and is interesting out of proportion to the 3.5 percent regional emissions it produces. Kazakhstan, Romania, and Uzbekistan each produce only about 4 percent of regional emissions and, with the exception of Kazakhstan, have not been deeply engaged in climate policy. Belarus, Hungary, Bulgaria, and all other countries in the region produce 2 percent or less of regional emissions. Bulgaria and Hungary have been engaged in climate discussions quite deeply, nonetheless.

There is a catch to price reform. Every increase in the price of gas, oil, or district heat increases the incentive to use low-cost, low-quality coal. In the Czech Republic, the majority of space heating is provided with coal. More important, perhaps, climate is simply not a priority for a distressed population. In a market democracy, priorities of the people are reflected by their government. The chief official concerns for the Czech government's energy policy are

- Security
- Affordable price
- Local environmental quality
- Entry into the European Union

The first and second priorities are not helpful for climatic change policy, but the last two are. There is a strong complementarity between cleaning up local air pollution and reducing greenhouse gas emissions. The Czech government provided over $0.5 billion in subsidies to clean up dirty coal facilities in the first half of 1996. The funding went primarily to pay the cost of physically switching plants from coal to natural gas.The Czech program for "healing the atmosphere" provides a prime example of the virtue of this approach. To reduce air pollution, the Czech government provided nonbusiness energy users with grants and businesses with zero-interest five-year loans. Forty percent of government assistance was provided as loans, 60 percent as subsidies, and the program was oversubscribed by half. Most effort was directed at replacing brown coal use. Savings in carbon emissions as a result of the program are expected to reach 2.4 million tons of carbons per year in 2000.

A study of the potential for co-benefits in Poland estimated that reducing sulfur dioxide and nitrogen oxides by 500,000 and 200,000 tons, respectively, would also reduce carbon dioxide emissions by 100 million tons annually by 2030.[48] These totals are equal to one-fifth the current sulfur dioxide and nitrogen oxide emissions in Poland.[49] Polish air pollution control was given a boost in 1991 when the Polish Sejm, or parliament, enacted its National Environmental Policy. Among its main goals are the reduction of air pollution, including particulates and sulfur, and reducing effluents into water. Poland has imposed a carbon tax, although it is small: $0.04 per ton of carbon dioxide. In addition, Poland has a well-developed environmental tax policy that contributes funds to a National Fund for Environmental Protection, which makes investments in efficiency and other environmental measures.

Even if one were to resolve the issues dividing the world at Kyoto, there remains the question of whether the transition economies have developed their legal and institutional infrastructures to the point where they would be able to participate in sophisticated emissions trading or joint implementation schemes. Recent research indicates that, without substantial efforts to develop the emissions baselines, inventorying, and verification measures necessary to implement international mitigation measures in Russia and Ukraine, those key nations will be unable to join in them, and may indeed fail to meet the basic requirements for participation in the Framework Convention on Climate Change. Researchers describe climate program leaders in those countries off-the-record as "fighting for their lives" and on the brink of failure.[50]

Conclusions

The transition economies represent probably the world's largest and cheapest greenhouse gas emissions mitigation opportunities. Cooperation to reduce emissions—to prevent regional emissions from returning to previously high levels—would serve both global and regional interests. Serious obstacles hinder the further development of climate protection programs in the transition economies. Regional governments place little priority on global climate. The region is already burdened with high expenditures to reduce sulfur dioxide, particulate, and water pollution. Most important, financing for mitigation measures—even cost-effective, profitable ones—is sorely lacking.

Overcoming these problems will be a challenge, particularly given the hostility of some nations to policies such as emissions trading. Even regional governments are not likely to press for such cooperation, as evidenced by lack of efforts to answer even the most fundamental questions about trading mechanisms, monitoring, and evaluation. How would emissions permits be allocated? How would emissions reductions be measured when emissions baselines are very difficult to establish?[51] These technical issues, as much as hostility from Europe and developing countries, may force the Framework Convention to reconsider the Kyoto approach, if the crucial participation of the transition economies in climate protection is to be obtained.

Improved cooperation on climate between the region and the west could help both the nations involved and the issue of climate protection. The most useful form of cooperation would come in the form of finance and infrastructure development. Financial backing is needed to capture low-cost emissions reductions in investments that have high paybacks but also high risk due to lack of credit history, collateral, and stability. Sometimes assistance can be most useful when it subsidizes the cost of project preparation because investors are wary—and increasingly so—of incurring up front costs for projects that may not be worthy of finance because of political risks. Infrastructure is needed to ensure that baselines for emissions trading on a project basis are developed fairly and honestly, and that verification and monitoring ensure that emissions reductions are real. Developing this infrastructure is, however, more in the interest of would-be purchasers of emissions credits—the West—than for the transition economies. Transition economy leaders hesitate to develop such infrastructure because they do not see a clear market for or commitment to developing a market for emissions reductions. Would-be buyers, of course, cannot very well buy emissions reductions until the infrastructure is in place to ensure the credibility of trading. Because the cheapest emissions reductions opportunities are probably in Russia and the formerly planned economies, creating the institutions for facilitating joint implementation or emissions trading is a priority.

Whether the Kyoto targets are met, additional measures beyond what today is cost effective are likely to be needed. Priority might be placed on sharing fundamental research for the development of new technologies to meet the climate challenge. Cooperation in developing fuel cells, advanced turbines, and renewable technologies could be very useful in the long term. Scientific collaboration could also help satisfy short-term security needs to support ex-Soviet science and to improve international relations that have been strained by economic hardships.

Notes

1. John H. Gibbons, *This Gifted Age: Science and Technology at the Millennium* (Woodbury, N.Y.: American Institute of Physics, 1997), 196.

2. By ratifying the FCCC, these nations accepted the same obligations as the other Annex I countries. The most important of these obligations is, "returning by the end of the present decade to earlier levels of anthropogenic emissions of carbon dioxide and other greenhouse gases not controlled by the Montreal Protocol" (Art. 4, paragraph 2a). The year 1990 was generally indicated as the reference year for emissions stabilization (Art. 4, paragraph 2b). The Countries with Economies in Transition were allowed by the Convention to apply "a certain degree of flexibility . . . with regard to the historical level of anthropogenic emission of greenhouse gases" (Art. 4, paragraph 6). Poland, for example, adopted 1988 as its reference year.

3. The carbon dioxide equivalents of these gases are measured against a 1995 baseline, not a 1990 baseline.

4. British Petroleum, *Annual Statistical Review, 1998*; BPAmoco, "Statistical Review of World Energy," http://www. bpamoco.com/worldenergy/.

5. Or an alternative year between 1989 and 1992.

6. Jae Edmonds, personal communication, Washington, D.C., 14 July 1988.

7. See also William Chandler, Meredydd Evans, and Alexander Kolesov, "Climate Change Mitigation: A Review of Cost Estimates and Methodologies for the Post-planned Economies." *Energy Policy* 24, no. 10 (1996): 927–935; and William Chandler, Alexei Makarov, and Zhou Dadi, "Energy for the Soviet Union, Eastern Europe," *Scientific American* 263 (October 1990): 121–127.

8. For an exhaustive description of what is now known about global warming, see Robert Watson, Marufu C. Zinyowera, and Richard H. Moss, eds., *Climate Change 1995: Impacts, Adaptations, and Mitigation of Climate Change: Scientific-Technical Analyses*, Contribution of Working Group II to the Second Assessment Report of the Intergovernmental Panel on Climate Change (Cambridge: Cambridge University Press, 1996).

9. John E. Oliver and Rhodes W. Fairbridge, eds., *The Encyclopedia of Climatology* (New York: Van Nostrand Reinhold, 1987), 185–195 and 660–670.

10. Watson et al., *Climate Change 1995*, 3–12.

11. Ibid., 22.

12. Oliver and Fairbridge, *Encyclopedia of Climatology*, 767–769.

13. Ibid., 691.

14. Watson et al., *Climate Change 1995* 229–230.

15. Ibid., 229–231, 257, 327–328, 442–443.

16. Ibid., 250.

17. Ibid., 254.

18. Peter Rutland, "Sakha (Yakutiya) Is Melting," *IEWS Russian Regional Report* 2, no. 26 (17 July 1997).

19. Watson et al., 250.

20. Ibid., 248.

21. Ibid., 103.

22. Ibid., 107–110.

23. Ibid., 119.

24. Sergei Krasnopeyev, Alexander Shevliakov, and Valery Kulikov, "Amur Tiger Conservation Plan Advances in Russian Far East" (Internet/mimeo provided by World Wildlife Fund, Russia, undated). The authors, respectively, are with Primorye Indigenous Peoples Association, Scientific Research Center, Vladivostok; Sikhote-Alin Reserve, Ternei; and Wildlife Foundation, Khaborovsk.

25. IPCC, Working Group II, 429.

26. Ibid., 432–433.

27. Ibid., 443.

28. Ibid., 446.

29. Ibid., 443.

30. Watson et al., 34.

31. Chandler, Evans, and Kolesov, "Climate Change Mitigation," 928.

32. Ibid.; A. S. Manne, *Global 2100: Alternative Scenarios for Reducing Emissions*, OECD Working Paper 111 (Paris, 1992).

33. Alexei Makarov and Igor Bashmakov, "The Soviet Union: A Strategy Development with Minimum Emission of Greenhouse Gases," *Energy Policy* 19, no. 10 (December 1991): 987–994.

34. Includes work underway at the Pacific Northwest National Laboratory, led by Jae Edmonds and Hugh Pitcher.

35. Jae Edmonds and John Reilly, *Global Energy: Assessing the Future* (New York: Oxford University Press, 1985), 279.

36. Henryk Gaj and Maciej Sadowski, *Climate Change Mitigation Case Studies from Poland*, edited by Susan Legro, Pacific Northwest National Laboratory (October 1997); www.pnl.gov/aisu/pubs.

37. Mykola Raptsun and Natalya Parasyuk, "Assessment of GHG Mitigation Technology Measures in Ukraine," *Applied Energy* 56, nos. 3/4 (1997): 367–380.

38. Aleksandr Avdiushin et al., *No-Regrets Options in Climate Change Mitigation Policy: Lessons from Transition Economies*, edited by Susan Legro, Battelle, Pacific Northwest National Laboratory (May 1997); B. Baranov, "Automation Control System and Electric Power Quality Check at Moscabelmet," *Energy Efficiency* 1, no. 3(1994): and I. Bashmakov, I Gritsevich, and S. Sorokina, *System of Institutional Measures to Stimulate Energy Efficiency in Russia as a Part of the GHG Mitigation Strategy* (Moscow: CENEf, 1996).

39. Jaroslav Marousek et al., *Climate Change Mitigation: Case Studies from the Czech Republic*, Pacific Northwest National Laboratory (28 February 1998); http://www.pnl.gov/aisu/pubs/czechcase.pdf.

40. Note that according to estimates using British Petroleum data, global energy demand in 1996 was an estimated 368 EJ, with carbon emissions totaling 6.4 billion tons.

41. British Petroleum, *Annual Statistical Review, 2000,* at www.bp.com.

42. See Janos Kornai, *The Economics of Shortage* (New York: Elsevier North-Holland, 1980); Chandler, Makarov, and Dadi, "Energy for the Soviet Union, Eastern Europe, and China," 121–127.

43. T M. Wigley, R. Richels, and J. A. Edmonds, "Economic and Environmental Choices in the Stabilization of Atmospheric CO2 Concentrations," *Nature* 379 (1996): 240–243.

44. Aleksandr Avdiushin et al., *No-Regrets Options in Climate Change Mitigation Policy: Lessons from Transition Economies,* Battelle, Pacific Northwest National Laboratories (May 1997).

45. The author shares this view.

46. The term "sinks" refers to an aspect of the natural environment that sequesters carbon or other greenhouse gases. Carbon sequestration is the biochemical process through which carbon in the atmosphere is absorbed by biomass such as trees, soils, and crops. The Kyoto Protocol endorsed the enhancement of national sinks–planting trees, for example, or increasing the carbon content of soils, for another–as a legitimate means of meeting greenhouse gas emissions reductions goals. The negotiators in Kyoto could not, however, agree on the method of measuring baselines of carbon contents in sinks or the rules by which countries could take credit for enhancing sinks, leaving that to subsequent meetings of the Conference of Parties of the Framework Convention on Climate Change.

47. Meredydd Evans, *Joint Implementation in Countries in Transition: An Analysis of the Potential and the Barriers,* PNNL–10480, Pacific Northwest National Laboratory (Richland, Wash., 1995); http://www.pnl.gov/aisu/pubs/jiforweb.htm; and Meredydd Evans, Susan Legro, and Ilya Popov, *The Climate for Joint Implementation: Case Studies from Russia, Ukraine, and Poland,* Pacific Northwest National Laboratory (27 October 1999); http://www.pnl.gov/aisu/pubs/jipaper.PDF.

48. Note that this reduction would apply to a much higher carbon dioxide emissions total than today's as a result of expected economic growth over the period.

49. Aleksandr Avdiushin et al., *No-Regrets Options in Climate Change Mitigation Policy.*

50. Susan Legro and Inne Gritcevich, *Climate Change Policy and Programs in Russia: An Institutional Assessment,* PNNL–13034, Battelle, Pacific Northwest National Laboratory (Washington, D.C., September 1999).

51. If no energy use meters existed previously, energy-related emissions can be estimated with difficulty.

8

The Second Generation
of Reform

An old joke in physics goes, "We have achieved full speed, but not momentum."[1] Although some of the transition economies are becoming stable market democracies, others appear to be going nowhere fast. The first decade of reform has been less successful than hoped because privatization and liberalization have outpaced institutional and legal development. Progress in the energy sector, in particular, has been impeded by issues of equity, rule of law, security, and pollution associated with providing heat, light, and work. A second generation of reformers, less constrained by ideology, could now apply practical experience to provide purposeful direction to the transition. This leadership can come from within the nations in transition but still needs the right kind of support from the international community.

End of the Beginning

The first phase of macroeconomic reform rewarded a few transition economies with recovery. Countries in the region with low deficits and high rates of private entrepreneurship have enjoyed economic growth and reduced inflation. The Czech Republic and Poland found the political will to impose austere budgets and severe restructuring. Along with Hungary, they have privatized their economies, brought inflation below 20 percent, and regained pre-1991 GDP levels.[2] Estonia, Latvia, and Lithuania measure well on these scales, as do Slovakia and Slovenia. Bulgaria and Romania long remained on the periphery of reform, but in recent years have also made progress. Belarus retains authoritarian controls, prints money, and shows deceptively positive statistical results. Russia, which owes $130 billion, ranks as the world's largest sovereign debtor, and both Russia and Ukraine—set back hard by the 1998 financial crisis—remain well below their potential.[3] Russian president Vladimir Putin alarmed reformers by attacking the press and by abolishing the independent environment ministry. The Caspian region remains poor and authoritarian, but Azerbaijan, Kazakhstan, and Uzbekistan have seen some economic growth, albeit from the lowest income levels in the former Soviet Union. Almost half of the twenty-six nations of the transition economies continue to have unsustainable budget deficits.

Energy use has mirrored economic change, with consumption falling by one-third. Regional energy intensity had long been several times that of the west, exaggerated by over-investment in heavy industry, distorted price signals, soft budget constraints, and technical stagnation. Energy per unit of GDP[4] went up significantly in Ukraine and Central Asia during the 1990s and was just beginning to improve in Russia when financial disaster struck in 1998. Significantly, Central Europe has proven that reform can decouple the region from its high intensity energy regime. Energy intensity in Poland, the Czech Republic, and Hungary has fallen by one-fifth.

Economist János Kornai, who best explained why central planning did not work, has proposed a set of ethical principles for the next phase of reform.[5] Kornai would increase the power of the individual and curtail that of the state, provide a social safety net, build a legal framework, and ensure transparency in government so that policy making is open to public view. Drawing on his long study of the region, he argues for balancing policy extremes, particularly sacrificing growth to pay for welfare against sacrificing welfare to encourage growth. To advocates of shock therapy he counsels patience, suggesting that "breakneck reform" allows no time for creating the institutions that are the foundations of market economies, so fundamental in the west that they are taken for granted, but so necessary in the transition economies to ensure lasting change (see Box 8.1).

Reforming Aid to Reform

Institutional deficiency has caused the energy sector to lag behind economic reform everywhere in the region. Energy prices have risen dramatically, but competition and transparency have not kept pace. Slow utility reform is perhaps understandable. After all, the United States' gargantuan utility, the Tennessee Valley Authority, is federally owned and suffers no competition. The French power sector has long been a model of central planning and an obstacle to European market integration. But lag is too small a word to describe the difference between potential and reality in transition economy energy issues. Natural gas, which could provide an excellent basis for transition, is nearly controlled by one company in one country. Nuclear power, which enjoys strong political support, operates without transparency, oversight, or the discipline of the

---

BOX 8.1   KORNAI'S GUIDELINES FOR REFORM

1. The Individual: Increase the power of the individual while curtailing that of the state.
2. Safety Net: Help the suffering and disadvantaged.
3. Competition: End state monopoly.
4. Efficiency: Encourage efficiency with incentives and ownership changes.
5. Law: Provide society with a legal framework, as the new role of the state.
6. Transparency: Make reform open to the public.
7. Patience: Allow time for lasting reform.
8. Balance: Compromise between sacrificing growth for welfare and welfare for growth.
9. Sustainable financing: Match government commitments to budgets that can be sustained.

SOURCE: Adapted from János Kornai, "Reform of the Welfare Sector in the Post-Communist Countries: A Normative Approach," in *Transforming Post-Communist Political Economies*, Joan M. Nelson, Charles Tilly, and Lee Walker, eds. (Washington, D.C.: National Academy Press, 1997), 272–298.

---

marketplace. District heating lacks market incentives for efficiency. Oil—vast amounts of it—has become a geopolitical tug-of-war, although this, of course, is better than the real war that could ensue. Leaders, to the extent that they pay attention to these matters, often concern themselves with shiny objects that satisfy short-term political needs. Governments inevitably deal more with expedience than principle, and so respond to the most pressing constituency, the most famous problem, the best photo opportunity. State dinners are for autocrats with oil, not consumer or environmental advocates. Russian-U.S. relations, moreover, have grown frostier as banking scandals and disputes over regional conflicts in Kosovo and Chechnya have spilled into the open.[6]

What does this criticism mean? Any objective measure would show that the oil and nuclear industries largely and disproportionately have benefited from local and western government support to the near exclusion of demand-side, consumer, and environmental matters. The amount of money spent on technical assistance; the number of loans from the World Bank, EBRD, and bilateral trade agencies; the number of regional government guarantees for internal and external loans; and even the number of press releases or conferences by subject matter weigh ten-to-one in favor of oil and nuclear over everything else. The public events themselves are often meant to satisfy a short-term need to produce a "de-

liverable" for a senior political figure's visit, to repay a favor, or to promote western business.[7]

One could make the case that this was only right, that oil production
is essential to earn hard currency, and oil is vital to western security.
Similarly, one could conclude that a repeat of Chernobyl represented a
profound threat, and preventing an accident took priority over all else.
One could further point to energy efficiency programs, such as the creation of centers of expertise with western funds, and structural adjustment lending to help ease unemployment in the coal sector, as evidence
that governments' response was more balanced. One could add that criticism is cheap and perhaps inevitably unfair to the players, who grapple
with competing demands for their time and resources. If analysts and
advocates fail to make a case for their own priorities, the fault may be
their own.

Another view is that the imbalance in money and leadership paid to oil
and nuclear power compared to the demand side, the consumer, and the
foundations of reform and innovation represents a grave mistake. No
one would dispute that the planned economies ranked among the most
energy inefficient in the world, and that consumers lacked the tools to respond to vastly higher prices, or even to measure the amount of heat they
actually consumed. Few could dispute that empowering consumers to
cut their costs and rationalize their consumption was a top priority—indeed essential—for reform to succeed. Yet, governments within the region and without failed to meet this challenge. One explanation for the
failure is that leaders were not sufficiently well-prepared. A less flattering
one is that moneyed lobbies usually set the agenda and altruistic notions
of public interest do not.

A U.S. presidential advisory panel, in contrast, recently concluded that
it was in the self-interest of the United States to pay more attention to the
foundations of energy innovation in transition and developing
economies (see Box 8.2).[8] The future of the global environment and world
economy—and therefore the well-being of Americans—depends on miti-
gating energy demand growth and the environmental impacts of energy
supply systems. The panel recommended that the U.S. president help
build local leadership capacity, support energy-sector reform in the transition economies and elsewhere, and demonstrate and help finance the
market penetration of energy efficient and environmentally benign energy technologies. Shifting to this brand of international energy cooperation, the panel argued, would provide more benefit to U.S. security,
trade, and environment than the general approach to technical assistance. A more-enlightened and productive approach would shift foreign
energy cooperation to aiding the individual and choosing reform rather
than aiding special interests or baldly seeking geopolitical advantage.

**BOX 8.2   FOUNDATIONS OF
ENERGY-TECHNOLOGY INNOVATION**

Capacity Building
• Increase support for regional technology and policy centers
• Expand existing and develop new training programs

Energy Sector Reform
• Provide assistance to 'get prices right' and advance public benefits
• Establish evolutionary regulatory frameworks for utilities
• Demonstrate and buy down cost of new technologies

Finance
• Shift multilateral banking transition to clean energy technologies
• Increase bilateral financing for energy efficiency and renewable energy

---

SOURCE: U.S. President's Committee of Advisors on Science and Technology, *Powerful Partnerships: The Federal Role in International Cooperation on Energy Innovation* (Washington, D.C.: White House Office of Science and Technology Policy, June 1999). The full text can be obtained at http://www2.whitehouse.gov/WH/EOP/OSTP/html/p2epage.html. The author was a committee member and author.

That the U.S., European, and some regional governments do have some experience with innovative and enlightened programs of energy cooperation should make it easier to effect the needed transition.

The Next Generation

The transition economies have multiple energy-related problems and must solve several of them at once. Both local leadership and foreign technical assistance are in short supply, making priority setting a kind of triage requiring careful matching of goal and intervention. Because the competition for technical assistance is a zero-sum game, priorities go begging, including economic recovery, jobs creation, municipal energy crises, climate emissions mitigation, and coal mine safety. Priority setting is inherently political, but to the extent that economic and technical analyses provide guidance, energy and environmental policy priorities might include creating a safety net, stopping nuclear proliferation, solving the utilities nonpayments problem, further downsizing the coal sector, writing the rules for utility competition, and rethinking western assistance.

Creating a Safety Net

High energy costs remain a problem for the entire region. Energy efficiency programs can help consumers cut costs and pollution and improve their economic prospects. Conservation programs targeted at financing improvements at industrial enterprises, heating utilities, and in the rapidly growing commercial sector make the most difference. So far, macroeconomic reform and industrial restructuring have contributed more to energy efficiency in the region than explicit energy-savings programs. Targeted energy efficiency programs now become increasingly important for continuing transformation and for equity and stability. Leadership in crafting and implementing these programs must come from within the region, but foreign assistance is needed. Interventions to promote efficiency have been most helpful when they have leveraged market interests and financial investment.

Countering Nuclear Proliferation

A response commensurate with the threat of nuclear weapons proliferation is needed to safeguard and oversee huge fissile material inventories in Russia. Plutonium production in that country could be halted with modest effort and nuclear materials reprocessing stopped, reducing two large flows of vulnerable weapons-grade material. Unfortunately, as thousands of nuclear weapons are being disassembled and their fissile materials made available for use in the Russian power industry, the Russian financial crisis destabilizes the establishment charged with overseeing and safeguarding all that critical material. Nuclear workers get paid poorly when at all, and drastic change and assistance for restructuring is needed if the Russian plutonium cities are to make the transition from weapons to benign commerce.

Requiring Payment for Service

The hemorrhage of money associated with the utility and energy supply nonpayments problem is a top priority for attention. Stanching these losses will require additional support for structural adjustment, regulatory reform, tariff collection, introduction of western accounting practices, and consumer assistance. Provision of "lifeline" utility rates and direct assistance to residential customers will be essential to restarting reform.

Building a Utility Regulatory Framework

The transition economies have experimented with competitive markets for electric power. Substituting competition for regulation is admirable and challenging but requires elementary regulation of power and fuel transmission and distribution facilities, consumer protection from monopoly control, and enforcement of the fundamental principle of fee-for-service. Without imposing the basic discipline that the consumer must pay for utilities, competitive utility models will never work in the region. Despite the crisis, the region can still afford to provide lifeline service to protect public health and safety, even as it requires payment for nonessential utility service.

Restructuring Coal

The death rate for coal miners in Ukraine is 100 times higher than in the United States. Sulfur and nitrogen oxide emissions in the region cause health and property damages totaling hundreds of dollars per ton of emission. New environmental control technology could make a difference in the region, but at the cost of using coal, although simple measures such as coal washing could partially and cheaply reduce air pollution. Increasing unemployment in the coal mining sector is inevitable. Continued financing of structural adjustment for Russia, Poland, and Ukraine to close unnecessary coal mines is sorely needed, in return for the region continuing the painful process of closing mines.

Rethinking International Cooperation

International energy cooperation can help solve problems critical to both east and west. Opening international energy cooperation to a broader community of business, environmental, and scientific interests can yield better ideas for achieving security, trade, and climate benefits. The first priority is creating the institutions that all western energy markets take for granted but that are essential for markets to work. The next priority is creative financing to adjust to higher prices, particularly on the demand side. A third priority is cooperation to create markets for natural gas as a transition fuel, particularly to secure access for Eastern Europe, South Asia, and China to prodigious gas supplies in Russia and the Caspian. A new priority is technology for western nations and the region jointly to develop and diffuse systems such as fuel cells that reduce the risk of climate change and the cost of emissions reduction. Taking a systems approach to these problems is needed, and there are signs that such an approach is actively being considered.[9]

* * *

Four hundred million people in the transition economies still struggle to leave behind the effects of planning, shortage, tyranny, Cold War, and war. Many will achieve the benefits of markets, opportunity, democracy, and cooperation. Their success, not only in the energy sector, will matter for those who would be their allies and for those who would mitigate new global threats.

Notes

1. Momentum, unlike speed, is defined by both velocity and direction. Thanks to physicist and presidential science advisor Jack Gibbons, who has sometimes made this observation.

2. Hungary's budget deficit, however, has remained consistently high. The Czech Republic slipped into recession in 1998–1999.

3. Arkady Ostrovsky and John Thornhill, "Russia 'Bust' Says Rating Agency," *Financial Times*, 24 September 1999, p. 3.

4. Measured in terms of estimated purchasing power parity.

5. János Kornai, "Reform of the Welfare Sector in the Post-Communist Countries: A Normative Approach," in *Transforming Post-Communist Political Economies,* ed. Joan M. Nelson, Charles Tilly, and Lee Walker (Washington, D.C.: National Academy Press, 1997), 272–298.

6. John Thornhill, Andrew Jack, Arkady Ostrovsky, and Phillip Jeune, "Russia 'Moved Reserves to Tax Haven," *Financial Times*, 12 February 1999, p. 1; Andrew Jack, John Thornhill, and Arkady Ostrovsky, "Revelations Rock Russian Central Bank," *Financial Times*, 12 February 1999, p. 2; Gerard Baker and John Thornhill, "Primakov Aborts U.S. Visit in Mid-Flight," *Financial Times*, 24 March 1999, p. 2; Andrew Jack, Jimmy Burns, and Charles Clover, "Yeltsin Family under Pressure as Financial Scandals Grow," *Financial Times*, 27 August 1999, p. 1; Andrew Jack, "The Tricks Russians Use to Funnel Money Abroad," *Financial Times*, 27 August 1999, p. 2; Thomas Catán, Jimmy Burns, and Arkady Ostrovsky, "The Quiet NY Bank in a Storm over Moscow Funds," *Financial Times*, 27 August 1999, p. 2.

7. An example of the second type was an effort to pacify interests in Kharkiv, Ukraine, who in 1998 gave in to pressure from the U.S. government to suspend sale of power turbines for a Russian-made nuclear reactor being built in Iran. In return, a "Kharkiv Initiative" was created to provide alternative economic opportunities for Kharkiv workers, particularly those in the "Turbatom" plant that manufactured the turbines and lost the sale. However, interviews with industry participants in and observers of the Kharkiv initiative complained confidentially that the initiative lacked funding. Similarly modest efforts can easily be recalled in association with multiple visits by U.S. secretaries of Energy, Commerce, and State, in both Republican and Democratic administrations. European agencies even more baldly engaged in activities that, to local experts, appeared to be transparent attempts to promote specific businesses or sectors in ways that were not consistent with the stated objective of providing aid. Creation of a European Union (EU) energy center in Bulgaria—and staffed by Greek expatriates who promoted EU businesses—is an example of trade subsidy substituting for aid.

8. U.S. President's Committee of Advisors on Science and Technology, *Powerful Partnerships: The Federal Role in International Cooperation on Energy Innovation*

(Washington, D.C.: The White House Office of Science and Technology Policy, June 1999). The full text can be obtained at http://www2.whitehouse.gov/WH/EOP/OSTP/html/p2epage.html. The author participated in this study.

9. Ambassador William Taylor, U.S. Department of State, personal communication, Washington, D.C., 25 April 2000.

Appendix:
Energy Conversions

Units to Be Converted

Barrels of oil	to	Joules	multiply by	6.1	[GJ/barrel]
Barrels of oil	to	BTU	multiply by	5.8	[MMBTU/barrel]
Barrels of oil	to	Gallons	multiply by	42.0	[Gallons/barrel]
Million barrels of oil/day	to	EJ/year	multiply by	2.20	[EJ/MMBD]
Tons of oil	to	Barrels	multiply by	7.00	[Barrels/ton]
Gallons of gasoline	to	Joules	multiply by	132,000	[kJ/gallon]
Gallons of gasoline	to	BTU	multiply by	125,000	[BTU/gallon]
BTU	to	Joules	multiply by	1,055	[Joules/BTU]
Kilocalories	to	BTU	multiply by	3.968	[BTU/kcalorie]
Kilocalories	to	Joules	multiply by	4,186	[Joules/kcalorie]
Cubic feet of gas	to	BTUs	multiply by	1,030	[BTU/cubic foot]
Cubic feet of gas	to	Joules	multiply by	1,087	[kJ/cubic foot]
Billion cubic meters	to	EJ	multiply by	1/26	[EJ/BCM]
Ton of coal	to	BTU	multiply by	22	[MMBTU/ton]
Kilowatt (electricity)	to	kWh/year	mutliply by	8.76E+03	[kWh/kW-year]
Megawatt (electricity)	to	kWh/year	multiply by	8.76E+06	[kWh/MW-year]
1000 MW (electricity)	to	kWh/year	multiply by	8.76E+09	[kWh/GW-year]
kWh	to	BTU	multiply by	3,412	[BTU/kWh]
kWh	to	Joules	multiply by	3.60E+06	[Joules/kWh]

Index

Printed in the United States
by Baker & Taylor Publisher Services

Printed in the United States
by Baker & Taylor Publisher Services